ENCYCLOPAEDIA OF
OPTICS

ENCYCLOPAEDIA OF OPTICS

Volume 1

Physical Optics

Mariyam Noor

ANMOL PUBLICATIONS PVT. LTD.

NEW DELHI - 110 002 (INDIA)

ANMOL PUBLICATIONS PVT. LTD.
4374/4B, Ansari Road, Daryaganj
New Delhi - 110 002
Ph.: 23261597, 23278000
Visit us at: www.anmolpublications.com

Encyclopaedia of Optics

First Published, 2006

ISBN 81-261-2613-2 (Set) 1

[Responsibility for the facts stated, opinions expressed, conclusions reached and plagiarism, if any, in this volume is entirely that of the Author/Editor. The Publishers bear no responsibility for them, whatsoever.]

PRINTED IN INDIA

Published by J.L. Kumar for Anmol Publications Pvt. Ltd., New Delhi - 110 002 and Printed at Mehra Offset Press, Delhi.

Contents

Preface

Science is the sermon of today. In every walk of life, since childhood till old age, science plays an important role in the life of man. Uptill some decades back, studied merely as a subject in schools and colleges, Science has assumed the status of an essential part of life today. With the expansion of literacy among the masses and growing awareness, they have woken up from the slumber of ignorance and have started taking up things more sincerely and thoughtfully. It is one of the reasons that our nation is heading towards new goals day after day.

It is a vast subject to study and talk about. For the purpose of ease, Science has been divided into more than two hundred and fifty branches. Specialisation in any branch can be obtained at different institutions and universities, scattered all over the country and abroad.

Apart from other super-specific branches, Physics, Chemistry, Zoology, Botany, Mathematics and Computronics, etc. are more general and closer to daily human life. From the point of view of the cut-throat competition in educational world, a thorough study of the above-mentioned branches, proves very handy, for the new aspirants, especially and for teachers and tutors, generally.

An urgent need was being felt for such a series of books, encompassing the commonly-studied and discussed branches of Science of equal significance and presented in an easy language, understood by all. With God's grace, our dream came true and we are happy, while introducing the dream-series for the lovers of knowledge.

The series consists of many a title, covering various branches of Science, such as Physics, Chemistry, Zoology, Botany, Geology, Agriculture Science, Computer Science, Mathematics and so on. The series is important and a must not only for students but also for teachers. Explained in simple language and aided with graphics and diagrams, everything, between the two ends—title and back cover—of the series, has been presented in a perfect way.

This set of books is devoted to **Optics**, which comprises three volumes: (1) Physical Optics, (2) Geometrical Optics and (3) Optics and Light.

It is hoped that the series would serve both, the students community and the academics, equally and will prove to be a boon for those, aspiring for newer heights. Nevertheless, any suggestions for further betterment of the endeavour will be highly appreciated and solicited.

Editor

1

Introduction

The branch of optics which deals with the theories of the nature of light, the phenomena of interference, diffraction, and polarisation, and the mechanism of its production, emission and propagation through space is called Physical Optics. In regard to the nature of light several theories have been proposed from time to time which , we shall discuss in due course. For our present purpose, however, we shall mainly take into account the wave theory as proposed by Huygens. Therefore, giving the explanation of the various optical phenomena on this theory it is essential to understand what a wave motion is, how it is produced and how it is propagated.

We are all familiar with the waves on the surface of water produced by dropping a stone into a pond. They are quite complex but they easily serve to illustrate important features present in any wave motion. As soon as the particles of water are displaced from their normal positions the restoring forces due to gravity and surface tension come into play. If the restoring force of gravity predominates we get waves, and if the surface tension force is large we get ripples. These restoring forces result from some sort of elastic connections between the adjacent particles of the medium, and being proportional to the displacement, the particles are set in simple harmonic motion. This motion is periodic in nature, that is, it repeats itself exactly after equal intervals of time. At the end of each interval * the particle finds itself in the same position and having the same velocity. These movements are handed on from particle to particle and a disturbance travels in the medium which is termed as wave motion.

Thus a wave motion is *a form of disturbance in a medium and is due to the repeated periodic movements of the particles of the medium about their mean positions, the disturbance being handed on from one particle to the next in a definite phase relationship.* It is one of the most important modes of the transference of energy from one place to another, such as sound energy, light energy, electromagnetic energy etc. This type of periodic movement of the particles is called simple harmonic motion (S.H.M.) and is very important in the wave theory of light. It permeates all branches of physics including atomic physics.

There are sound waves, seismic waves, water waves, light waves, electromagnetic waves, and de Broglie matter waves. In this treatise we shall discuss light waves only in adequate details, but give a brief description of water waves, electromagnetic waves, de Broglie matter waves and atomic waves.

The displacement of the particles in S.H.M. is given by

$$y = a \sin \omega t, \qquad ...(1)$$

assuming that $y = 0$ when $t = 0$, or more generally

$$y = a \sin (\omega t \pm \alpha) \qquad ...(2)$$

where α is the epoch or initial Phase.

Type of waves. There are two main types of waves, (*a*) Longitudinal, (*b*) Transverse.

(*a*) *The longitudinal* waves are those in which the generating particles or the layers of the particles of the medium oscillate to and fro about their mean positions in line with the direction in which the wave traverses just as sound waves. They are due to pressure variations in the medium *i.e.*, excess and deficit of pressure forming condensations and rarefactions which travel on and on, while the medium remains stationary.

(*b*) The *transverse* waves are those in which the particles of the medium oscillate up and down at right angles to the direction in which the wave travels, the medium, however, remaining stationary, just as ripples on the surface of water. The light and electromagnetic waves are transverse, the former being of much shorter wavelength than the latter.

Equation of a simple harmonic wave. We have given the displacement equation of a single particle executing simple harmonic motion at any instant. Let us now consider a row of such particles in an elastic medium, along which a wave travels with a velocity v ; then we can find an equation which gives us the displacement of all the particles at a particular instant, as also the displacement of a single particle at any time. Such an equation will represent a simple harmonic wave.

Let O be the first particle in the row executing simple harmonic motion. Its displacement at any instant t is given by

$$y = a \sin \omega t$$

$$= a \sin \frac{2\pi}{T} t_2 \quad \ldots \textit{since } \omega = \frac{2\pi}{T}$$

where T is the periodic time.

Obviously, the displacement varies with time. As the wave same time period T but their displacements at any given instant will generally not be the same as that of the first particle. Consider a particle P distant x from the reference Particle O. If v be the velocity of propagation of the wave, then the particle P starts vibrating some time later than O given by $t_1=x/v$ and its displacement at any instant t would be the same as that of O, t_1 seconds earlier and accordingly it is represented by

$$y = a \sin \frac{2\pi}{T} (t - t_1)$$

$$= a \sin \frac{2\pi}{T}\left(t - \frac{x}{v}\right)$$

$$= a \sin 2\pi \left(\frac{t}{T} - \frac{x}{vT}\right)$$

$$= a \sin 2\pi \left(\frac{t}{T} - \frac{x}{\lambda}\right) \quad ...(3)$$

This equation can also be put as

$$y = a \sin \frac{2\pi}{\lambda}\left(\frac{\lambda}{T}t - x\right)$$

$$= a \sin \frac{2\pi}{\lambda} (vt - x) \quad ...(4)$$

The expression $2\pi\left(\frac{t}{T} - \frac{x}{\lambda}\right)$ in equation (3) is known as the phase angle and $\frac{2\pi x}{\lambda}$ gives the phase difference between the particles O and P, P being behind O in phase by $\frac{2\pi x}{\lambda}$. For $x = \lambda$, the phase difference is 2π. The same is true for $x = 2\lambda$, 3λ, ...

If instead of one there are two such particles P_1 and P_2 distant x_1 and x_2 respectively from the reference particle O, their equations of displacement will be

$$y_1 = a \sin \frac{2\pi x}{\lambda} (vt - x_1)$$

and

$$y_2 = a \sin \frac{2\pi}{\lambda} (vt - x_2).$$

Their phase difference

$$= \frac{2\pi}{\lambda}(vt - x_1) - \frac{2\pi}{\lambda}(vt - x_2)$$

$$= \frac{2\pi}{\lambda}(v_2 - x_1)$$

$$= \frac{2\pi}{\lambda} \times \text{(path difference)} \quad ...(5)$$

It is clear that the equations (3) and (4) above represent the equations of motion of a particle which is at a distance x from the reference particle O and starts its motion at a time t_1 later. Evidently any particular particle is specified by one value of x, and as the wave travels with a velocity v, the particles move up and down with a displacement determined by the ordinate of the curve, so that the equation of motion of any other particle can be obtained by giving x a particular value in the above equations. Thus the motions of all the particles are identical except for the progressive decrease if phase difference with the increase in x. Hence the equations (3) and (4) represent the equations of the wave motion travelling with a velocity v, having a wavelength A and amplitude a. Here the direction of wave motion represented by these equations is from left to right and for a wave travelling from right to left the sign of x changes, *i.e.*, the equations become

$$y = a \sin 2\pi \left(\frac{t}{T} + \frac{x}{\lambda}\right) \quad ...(6)$$

$$y = a \sin \frac{2\pi}{\lambda}(vt + x) \quad ...(7)$$

It may be noted that in a simple harmonic wave the displacement of the particles of the medium changes periodically both with respect to time and space.

Average kinetic energy of a particle in S.H.M. Consider the general equation of S.H.M.

$$y = a \sin (\omega + \alpha)$$

then velocity of the particle, $v = \frac{dy}{dt} = \alpha\omega \cos(\omega\tau + \alpha)$.

The kinetic energy of a vibrating particle of mass m and velocity v at any instant is given by $\frac{1}{2} mv^2$.

Since the velocity of the particle varies with time, its average kinetic energy E during a complete period of vibration is given by

$$E = \frac{1}{T}\int_0^T \frac{1}{2} mv^2 \, dt$$

$$= \frac{1}{T}\int_0^T \frac{1}{2} ma^2 w^2 \cos^2(\omega t + \alpha)\, dt$$

$$= \frac{ma^2w^2}{4T}\int_0^T 2\cos^2(\omega t + \alpha)\, dt$$

$$= \frac{ma^2w^2}{4T}\int_0^T [1 + \cos 2(\omega t + \alpha)]\, dt$$

$$= \frac{ma^2w^2}{4T}[t]_0^T \; ....$$

$$\left[\because \int_0^T \cos 2(\omega t + \alpha)\, dt = 0\right.$$

$$= \frac{ma^2\omega^2}{4T}[T] = \frac{ma^2\omega^2}{4}$$

$$= \frac{ma^2(2\pi/T)^2}{4} = \left[\frac{ma^2\pi^2}{T^2}\right]$$

$$= ma^2\pi^2n^2 \qquad ... \left[\because \frac{1}{T^2} = n^2\right]$$

$$\therefore \quad E = \mathbf{ma^2\pi^2n^2} \qquad ... (8)$$

Thus, the average kinetic energy of a particle vibrating in S.H.M. is directly proportional to the square of its amplitude.

Composition of two or more colliaear S.H.Ms of equal periods but differing in phases and amplitudes. If a particle is executing at the same time two S.H.Ms in the *same straight line* and of the *same frequency* but of *different amplitudes* and *phases*, its resultant motion can be found either analytically or graphically.

(a) *Analytical method.* Let the two S.H.Ms be represented by

$$y_1 = a_1 \sin (\omega t + \alpha_1), \qquad ...(i)$$

$$y_2 = a_2 \sin (\omega t + \alpha_2). \qquad ...(ii)$$

The phase angle of first S.H.M. is $(\omega t + \alpha_1)$ while that of the second is $(\omega t + \alpha_2)$, so that $(\alpha_2 + \alpha_1)$ is the phase difference of the two. The resultant displacement

$$y = y_1 + y_2 = a_1 \sin (\omega t + \alpha_1) + a_2 \sin (\omega t + \alpha_2)$$

$$= \sin \omega t\, (a_1 \cos \alpha_1 + \alpha_2 \cos \alpha_2) + \cos \omega t\, (a_1 \sin \alpha_1 + a_2 \sin \alpha_2) \qquad ...(iii)$$

Let as choose an angle θ such that

$$a_1 \cos \alpha_1 + a_2 \cos \alpha_2 = R \cos \theta \qquad ...(iv)$$

and $$a_1 \sin \alpha_1 + a_2 \sin \alpha_2 = R \sin \theta, \qquad ...(v)$$

$$y = R\,(\sin \omega t \cos \theta + \omega t \sin \theta)$$

$$= R \sin (\omega t + \theta) \qquad ...(9)$$

which is the equation of an S.H.M. of the *same* time period but of phase $(\omega t + \theta)$ and amplitude R.

The values of R and θ can be obtained from *(iv)* and *(v)*.

Squaring and adding

$$*R^2 = (a_1 \cos \alpha_1 + a_2 \cos \alpha_2)^2 + (a_1 \sin \alpha_1 + a_2 \sin \alpha_2)^2$$

$$= a_1^2 + a_2^2 + 2a_1 a_2 \cos (\alpha_2 - \alpha_1),$$

which $$R = \sqrt{a_1^2 + a_2^2 + 2a_1\, a_2 \cos\left(\alpha_2 - \alpha_1\right)}, \qquad ...(10)$$

and $$q = \tan^{-1}\left(\frac{a_1 \sin \alpha_1 + a_2 \sin \alpha_2}{a_1 \cos \alpha_1 + a_2 \cos \alpha_2}\right). \qquad ....(11)$$

Special cases.

(*i*) If the two S.H. vibrations are in the same phase, *i.e.*, $\alpha_2 - \alpha_1 = 0, 2\pi, 4\pi$ etc., we have

$$\cos(\alpha_2 - \alpha_1) = 1$$

$$R^2 = a_1^2 + a_2^2 + 2a_1 a_2 = (a_1 + a_2)^2,$$

$$R = a_1 + a_2, \text{ which is maximum}$$

Thus the particle performs the S.H.M. with an amplitude equal to the sum of the respective amplitudes.

(*ii*) If the two S.H. vibrations are a quadrature apart, *i.e.*,

$$a_2 - a_1 = \frac{\pi}{2}, \frac{3\pi}{2}, \frac{5\pi}{2}, \text{ etc., we have}$$

$$\cos(\alpha_2 - \alpha_1) = 0$$

$$\therefore \quad R^2 = a_1^2 + a_2^2,$$

or

$$R = \sqrt{a_1^2 + a_2^2}$$

(*iii*) When the two S.H. vibrations are in the opposite phase,

i.e., $\alpha_2 - \alpha_1 = \pi, 3\pi, 5\pi,$ *etc.*, we have

$$\cos(\alpha_2 - \alpha_1) = -1,$$

$$R = a_1 \sim a_2, \text{ which is minimum.}$$

Thus the resultant vibration of the particle will occur with minimum amplitude if the two S.H.Ms are in opposite phase.

The method can be extended to a large number of S.H. vibrations.

(*b*) *Graphical method.* Sometimes the graphical methods making use of the extension of the polygon law are very convenient for finding the resultant of a number of S.H. vibrations. We shall apply this method to one case of special importance in the theory of light.

Resultant of n simple collinear harmonic motions having equal amplitudes, same period and phases in increasing arithmetical progression. Let a be the constant amplitude and δ the common phase difference between successive components of all the n vibrations of the same period which act on a particle simultaneously. To find the amplitude and the initial phase of the resultant we proceed graphically to construct the polygon of amplitudes with sides $OA, AB, ...PQ,$ each equal to a. The closing side OQ of the polygon represents the resultant amplitude R and the angle θ which this makes with the first vibration OA gives the initial phase of the resultant.

Resolving along OX and OY, *viz*, parallel and perpendicular to first side OA ; we have

$$R \cos \theta = a \{ 1 + \cos \delta + \cos + 2\delta + \dots \dots + \cos (n-1) \delta\} \quad ...(i)$$

$$R \sin \theta = a \{ 0 + \sin \delta + \sin + 2\delta + \dots \dots + \sin (n-1) \delta\} \quad ...(ii)$$

Multiplying (*i*) by 2 sin ($\delta/2$), we get

$$2R \cos \theta \sin \frac{\delta}{2}$$

$$= a \left[2 \sin \frac{\delta}{2} + 2 \cos \delta \sin \frac{\delta}{2} + 2 \cos 2\delta \sin \frac{\delta}{2} + \dots\dots + 2 \cos (n-1) \delta \sin \frac{\delta}{2} \right].$$

Making use of the trigonometrical relation

$$2 \cos A \sin B = \sin (A + B) - \sin (A - B), \text{ we get}$$

$$2 R \cos \theta \sin \frac{\delta}{2}$$

$$= a \left[2 \sin \frac{\delta}{2} + \left(\sin \frac{3\delta}{2} - \sin \frac{\delta}{2} \right) + \left(\sin \frac{5\delta}{2} - \sin \frac{3\delta}{2} \right) + \dots \right.$$

$$\left. \dots + \left\{ \sin \left(n - \frac{1}{2} \right) \delta - \sin \left(n - \frac{3}{2} \right) \delta \right\} \right]$$

$$= a \left[\sin \frac{\delta}{2} - \sin \left(n - \frac{1}{2} \right) \delta \right]$$

$$= 2a \sin \frac{n\delta}{2} \cos (n-1) \frac{\delta}{2},$$

or $$R \cos \theta = \frac{a \sin (n\delta / 2)}{\sin(\delta / 2)} \cos (n - 1) \frac{\delta}{2} \qquad ...(iii)$$

Similarly, multiplying equation (*ii*) By 2 sin (δ/2) and simplifying, we get

$$R \sin \theta = \frac{a \sin(n\delta / 2)}{\sin (\delta / 2)} \sin(n - 1) \frac{\delta}{2}. \qquad ...(iv)$$

Squaring (*iii*) and (*iv*) and adding, we get

$$R^2 = \frac{a^2 \sin^2 (n\delta / 2)}{\sin^2 (\delta / 2)}$$

or $$R = a \frac{\sin (n\delta / 2)}{\sin (\delta / 2)} \text{ or } -a \frac{\sin (n\delta / 2)}{\sin (\delta / 2)} \qquad ...(12)$$

dividing (*iv*) by (*iii*), we get

$$\tan \theta = \tan \frac{(n - 1) \delta}{2}$$

or $$\theta = \frac{(n - 1) \delta}{2} \text{ or } \frac{(n - 1) \delta}{2} + \pi. \qquad ...(13)$$

If we take the second value of θ, consideration of the limiting case when n = 1 shows that we must then choose negative sign in taking the root of R^2 and consequently get the same result over again.

Special case. This arises when n is very large and both the amplitude a and the constant initial phase difference δ are intinitesimally small so that na and $n\delta$ remain finite The vibration polygon of then coincides with its circumscribed circle and the resultant is represented in amplitude and in phase by the chord *OQ* of the circular arc *OPQ*. The length of the arc *OPQ* is na, the sum of the amplitudes.

The phase of the resultant vibration is given by

$$\theta = \phi/2$$

where ϕ is the angle subtended by the chord *OQ* at the centre.

Now $$\phi = \frac{are\ OPQ}{radius\ CO} = \frac{na}{r}$$

$$\therefore \quad \theta = \phi/2 = na/2r. \quad \text{... (14)}$$

Then resultant, $$R = OQ = 2r \sin\frac{\phi}{2}\ 2r \sin\theta$$

$$= na\frac{\sin\theta}{\theta}\ \ldots\ldots \qquad \left[\because 2r = \frac{na}{\theta}\ by\ (14)\right]$$

When all the simple harmonic vibrations are in the same phase then the resultant magnitude

$$R_0 = \Sigma a = na$$

This result is of great importance and will be used later in the theory of the grating.

Composition of two rectangular S.H.Ms of equal periods but differing in phases and amplitudes. Let a particle be simultaneously subjected to two S.H.Ms having the same period but different amplitudes a and b, and the phases $(\omega t + \delta)$ and ωt, one acting along X-axis and the other along Y-axis and represented by the equations

$$x = a \sin(\omega t + \delta) \quad \text{...}(i)$$

and $$y = b \sin\omega t \quad \text{...}(ii)$$

where x and y are their displacements and δ is the phase difference between them.

To obtain the equation of the path in which the resultant motion of the particle will occur, we eliminate ω from (i) and (ii).

From (i), $$\frac{x}{a} = \sin\omega t\cos\delta + \cos\omega t\sin\delta$$

$$= \frac{y}{b}\cos\delta + \sqrt{1 - \frac{y^2}{b^2}}\ \sin\ \delta,$$

Since from (*ii*),

$$\sin \omega t = \frac{y}{b} \text{ and } \cos \omega t = \sqrt{1 - \frac{y^2}{b^2}}.$$

$$\therefore \quad \frac{x}{a} - \frac{y}{b} \cos \delta = \sqrt{1 - \frac{y^2}{b^2}}. \sin \delta.$$

Squaring both sides, we get

$$\frac{x^2}{a^2} + \frac{y^2}{b^2} \cos^2 \delta - \frac{2xy}{ab} \cos \delta = \left(1 - \frac{y^2}{b^2}\right) \sin^2 \delta$$

or
$$\frac{x^2}{a^2} + \frac{y^2}{b^2} (\cos^2 \delta + \sin^2 \delta) - \frac{2xy}{ab} \cos \delta = \sin^2 \delta,$$

or
$$\frac{x^2}{a^2} + \frac{y^2}{b^2} - \frac{2xy}{ab} \cos \delta = \sin^2 \delta$$

which is the general equation of an ellipse with its major and minor axes inclined to the coordinate axes at a certain angle depending upon their relative phases.

Hence the resultant motion of the particle occurs in an elliptic path.

Special cases.

Case I. When the phase difference between the two S.H.Ms is zero, *i.e.*, $\delta = 0, 2\pi......2n\pi$, the equation of the path of the resultant motion is given by

$$\frac{x^2}{a^2} + \frac{y^2}{b^2} - \frac{2xy}{ab} = 0,$$

or
$$\left(\frac{x}{a} - \frac{y}{b}\right)^2 = 0, \text{ or } \pm \left(\frac{x}{a} - \frac{y}{b}\right) = 0 \qquad ...(17)$$

This is the equation of two coincident straight lines and shows that the resultant vibrations of the particle occur in a straight line passing through the origin of the rectangular axes and making an angle, $\theta = \tan^{-1} (b/a)$ with the positive direction of the X-axis. It is noteworthy that in this case elliptic vibrations have shrunk into linear vibrations.

Also when $a = b$, the angle θ will be 45° and the resultant vibrations of the particle are along a straight line passing through the origin and inclined to the positive direction of the X-axis at an angle of 45°.

Case II. When the phase difference between the two S.H.Ms is $\delta = \pi, 3\pi \ldots\ldots (2n + 1)\pi$, the equation of the path of the resultant motion of the particle is given by

$$\frac{x^2}{a^2} + \frac{y^2}{b^2} + \frac{2xy}{ab} = 0,$$

or $$\left(\frac{x}{a} + \frac{y}{b}\right)^2 = 0$$

or $$\pm\left(\frac{x}{a} + \frac{y}{b}\right) = 0 \qquad \ldots (18)$$

which again is the equation of two coincident straight lines passing through the origin and shows that the resultant vibrations of the particle again occur in a straight line passing through the origin of the rectangular axes and making an angle, $\theta = tan^{-1}(-b/a)$ with the *negative* direction of the X-axis.

Also when a and b are equal, θ =45° and the resultant vibrations are along a straight line passing through the origin and making an angle of 45° with the negative direction of the X-axis.

The above analysis shows that if two S.H.Ms about a common centre with equal periods and in directions mutually perpendicular, be compounded the resultant vibration is a S.H. motion if $\delta = 2n\pi$. But, if $\delta = (2n + 1)\pi$, the resultant vibration while still simple harmonic is rotated through an angle 2θ, where $\theta = \tan^{-1}(b/a)$ if θ be the angle the first vibration makes with either axis.

This statement is of importance in the theory of polarized light.

Case III. (*a*) When the phase difference the equation of motion of the particle reduces to

$$\frac{x^2}{a^2} + \frac{y^2}{b^2} = 1. \qquad \text{... (19)}$$

which is the equation of a symmetrical ellipse, which shows that the resultant motion of the particle occurs along an elliptic path with major and minor axes coinciding with the rectangular axes along which the two S.H.Ms are occurring.

(*b*) If in addition to a phase difference $\delta = \pi/2,\ 3\pi/2$ etc., the amplitudes of both the S.H.Ms are equal, then the equation of the resultant path reduces to

$$\frac{x^2}{a^2} + \frac{y^2}{b^2} = 1,$$

$$x^2 + y_2 = a^2, \qquad \text{... (20)}$$

which is the equation of a circle with radius a and origin at O.

Thus, if a particle is subjected to two rectangular. S.H.Ms of equal periods and amplitudes but differing in phase by $(2n + l)\pi / 2$, *it moves uniformly in a circular path.*

Conversely, *a uniform circular motion can be resolved into two rectangular linear S.H.Ms of equal periods and amplitudes but differing in phase by* $(2n + l)n/2$.

(*c*) If there is a slight difference in periodic times of the two vibrations, the elliptic paths described above will not remain steady but will undergo changes.

The changes in the resultant path as the phase difference increases from 0 to π. As the phase difference changes from π to 2π.

2

Matter Waves

The quantum hypothesis of radiation provided both qualitative and quantitative explanation of 'black body' radiation (1901), photoelectric effect (1905), emission of line spectra (1913) and Compton effect (1924), etc. On the other hand, the electromagnetic (Wave) theory gave a satisfactory account of the phenomena of interference, diffraction and polarisation ; apparently inexplicable by the former. Thus the experiments established beyond doubt the *dual aspect* of radiation *(wave aspect* and *particle aspect).*

But despite the immense success of wave-particle duality of radiation, the two fundamental postulates underlying Bohr's theory of atomic structure remained arbitrary and unexplained for a long time and agitated intuitive minds. In the mean time the Einstein's mass-energy relationship, $E = mc^2$, had been verified in a number of cases, according to which radiation and mass were mutually convertible. It was in this background, that the French scientist Louis de Broglie in 1924 extended the idea of dual nature of radiation to matter itself despite their distinguishing features, when he proposed that matter possesses wave as well as particle characteristics and that motion of electrons within the atom is guided by peculiar kind of waves which were called 'pilot waves', in the same manner as the radiation field governed by electromagnetic theory served to guide the photons. According to his hypothesis not only electrons but all material particles like protons, neutrons, atoms or molecules have each a wave associated with them, which are called 'matter waves'.

Expression for the de Broglie wavelength. On the quantum theory of radiation, the energy of a photon of frequency ν is given by

$$E = h\nu. \qquad cf \text{ Eq. (23-1)}$$

According to Einstein's mass-energy relationship ($E = mc^2$) the photon may be considered to be endowed with a mass m given by

$$m = E/c^2 = h\nu/c^2$$

Therefore, the momentum of the photon

$$p = mc = \frac{h\nu}{c^2} \times c = \frac{h\nu}{c} = \frac{h}{\lambda} \ldots \left[\because \nu = \frac{c}{\lambda}\right]$$

or

$$\lambda = \frac{h}{mc} \qquad \ldots(1)$$

where λ is the wavelength of the photon having momentum mc.

It was by analogy with this equation associating momentum with a photon that de Broglie proposed the concept of matter waves, according to which a material particle of mass m moving with a velocity v should have an associated wavelength λ, called the de Broglie wavelength, given by

$$\lambda = \frac{h}{\text{momentum}} = \frac{h}{mv} \qquad \ldots (2)$$

where h is Planck's constant and m the relativistic mass of the particle given by

$$m = \frac{m_0}{\sqrt{1 - \frac{v^2}{c^2}}}$$

m_0 is the rest mass, v the velocity of the particle and c the velocity of light. Equation (2) is known as the de Broglie wave equation. This equation combines the wave particle duality concepts—*corpuscular* as well as *undulatory*, in a very intimate way, for λ has clean-cut meaning only in connection with a *wave* and mv, the momentum with a *moving particle.*

Wave velocity and group velocity for de Broglie waves. According to de Broglie's view each particle of matter such as electron, proton and neutron etc. may be regarded as consisting of a *group* of waves or a wave packet. Each component wave propagates with a definite velocity called the wave velocity or phase velocity. But when a disturbance consists of a number of component waves, each travelling with a slightly different velocity, the resultant (measured) velocity will be that of a *periodicity* (or modulation) impressed upon the wave packet. *The velocity of advance of this periodicity is called the* group velocity. It is denoted by v_g. If we denote de Broglie's phase velocity of a particle, by v_ϕ then by the usual formula

$$v_\phi = \nu\lambda$$

Now $\lambda = h/mv$, where v is the particle velocity and m its relativistic mass. By photon analogy $\nu = E/h = mc^2/h$, since $E = mc^2$ (Einstein),

The de Broglie phase velocity is, therefore,

$$v_\phi = \nu\lambda$$

$$= \frac{mc^2}{h} \times \frac{h}{mv} = \frac{c^2}{v} \qquad \text{... (3)}$$

Since the velocity v of a material particle cannot be greater than c, but the de Broglie phase velocity v_ϕ is always greater than c. Obviously, this is not the velocity with which the de Broglie wave packet associated with a particle travels. The phase velocity of de Broglie waves has no simple significance in itself.

To find the group velocity v, we first find the angular frequency ω and the propagation constant k of the de Broglie wave associated with a body travelling with a velocity v.

The angular frequency ω of the de Broglie wave is

$$\omega = 2\pi\nu = \frac{2\pi mc^2}{h}$$

since by photon analogy

$$\nu = \frac{mc^2}{h}$$

or

$$\omega = \frac{2\pi m_0 c^2}{h\left(1 - v^2/c^2\right)^{\frac{1}{2}}}$$

$$\ldots \left[\because m = \frac{m_0}{\left(1 - v^2/c^2\right)^{\frac{1}{2}}}\right] \quad \ldots (4)$$

The propagation constant k is,

$$k = \frac{2\pi}{\lambda} = \frac{2\pi mv}{h} = \frac{2\pi m_0 v}{h\left(1 - v^2/c^2\right)^{\frac{1}{2}}} \quad \ldots (5)$$

The de Broglie group velocity

$$v_0 = \frac{d\omega}{dk} = \frac{d\omega}{dv} \div \frac{dk}{dv}$$

Differentiating Eq. (4) and (5) with respect to v, gives

$$\frac{d\omega}{dv} = \frac{2\pi m_0 v}{h\left(1 - v^2/c^2\right)^{\frac{3}{2}}} \quad \ldots (5\,a)$$

and

$$\frac{dk}{dv} = \frac{2\pi m_0}{h\left(1 - v^2/c^2\right)^{\frac{3}{2}}} \quad \ldots (5b)$$

$$\therefore \qquad v_g = \frac{\frac{d\omega}{dv}}{\frac{dk}{dv}} = \frac{2\pi m_0 v}{h\left(1-\frac{v^2}{c^2}\right)^{\frac{3}{2}}} \div \frac{2\pi m_0}{h\left(1-\frac{v^2}{c^2}\right)^{\frac{3}{2}}} = v$$

i.e., the de Broglie wave packet travels with the body.

Nature and properties of the de Broglie matter waves

(*i*) The component waves of the de Broglie wave packet have amplitudes which vary with the *likelihood* of detecting the particle at a particular position as depicted in the schematic diagram of the wave packet.

The de Broglie waves are, therefore, termed as probability waves or "Pilot waves" as their function is also to guide the motion of the particle.

(*ii*) The wavelength of the de Broglie waves depends upon the mass of the particle, since $\lambda = h/mv$. Consequently greater the mass, the smaller the wavelength. The electrons, protons and neutrons travelling with the same velocity will have different wavelengths.

(*iii*) The phase velocity v_ϕ of the de Broglie waves is greater than that of electromagnetic waves $\left(v_\phi = \frac{c^2}{v}\right)$.

(*iv*) The de Broglie waves are quite unlike the electromagnetic waves which travel with a constant velocity c (= 3×10^2 ms^{-1}) as their phase velocity v_ϕ is variable depending as it does on the mass and the velocity of the material particles.

The de Broglie Electron Wave

From Equation (3) we find that if the particles, say electrons are accelerated to various velocities, we can produce waves of various lengths. The higher the electron velocity, the smaller the "de Broglie wavelength". If the velocity v is given to an electron by accelerating it through a potential difference V then the work done on the electron is Ve. This work is converted into the kinetic energy K of the electron, thus

$$K = \tfrac{1}{2}mv^2 = Ve \qquad \text{... (6)}$$

$$v = \sqrt{\frac{2Ve}{m}}$$

or momentum, have

$$mv = \sqrt{2mVe} \qquad \text{... (7)}$$

Substituting this value in the de Broglie equation, $\lambda = \dfrac{h}{mv}$, we have

$$\lambda = \frac{h}{\sqrt{2mVe}} \qquad \text{... (8)}$$

Ignoring relativistic considerations, $m = m_0$, the rest mass

$$\therefore \qquad \lambda = \frac{h}{\sqrt{2m_0Ve}} \qquad \text{... (9)}$$

Now in SI system,

$$h = 6.625 \times 10^{-24} \text{ J-s.}$$

$$m_0 = 9.11 \times 10^{-31} \text{ kg}$$

$$e = 1.6 \times 10^{-19} \text{ C.}$$

$$\therefore \qquad \lambda = \frac{h}{\sqrt{2m_0Ve}}$$

$$= \frac{6.625 \times 10^{-34} \text{ J-s}}{\sqrt{2 \times 9.11 \times 10^{-31} \text{ kg} \times 1.6 \times 10^{-19} \text{ CV}}}$$

$$= \frac{12.27}{\sqrt{V}} \times 10^{-10} \text{ m} = \frac{12.27}{\sqrt{V}} \text{ Å} \qquad \text{... (10)}$$

where V *is the numerical value of p.d. in volts.*

Note that 150 volts are required to produce electrons of wavelength 1 Å, in contrast with *X*-rays which require approximately 12,000 volts for 1 Å. The larger, therefore, is the velocity of the electron, the smaller is the de Broglie wavelength. Davisson and Germer in their original experiments on electron diffraction used electrons of 20–600 electron-volts.

Modern diffraction equipment employs usually voltages of the order of 50 kilovolts, corresponding to a wavelength of about 0.05A. In such a case a relativistic correction on account of variation of mass with velocity must be applied to equation (3). It can easily be seen that for 50 kev electrons this correction lowers the wavelength by approximately 2.5%, see equation (11). Electron generated de Broglie waves are of very short wavelengths and are quite useful for certain types of diffraction analysis.

Relativistic Correction in the Formula

Necessary correction in equation (10) on account of the relativistic increase in the mass of the high velocity electrons may be made, since the mass according to the theory of relativity is not an invariable entity as in Newtonian physics, but is given by

$$m = \frac{m_0}{\left(1 - \frac{v^2}{c^2}\right)^{\frac{1}{2}}},$$

where m_0 is the rest mass of the particle, v its velocity and c the velocity of light.

Applying relativistic equation (25-30) for kinetic energy E_k, *viz.*,

$$E_k = (m - m_0)c^2, \text{ we have}$$

$$(m - m_0)\, c^2 = Ve$$

$$\text{or} \qquad m - m_0 = \frac{Ve}{c^2}$$

$$= m_0 + \frac{Ve}{c^2} = m_0\left(1 + \frac{Ve}{m_0c^2}\right)$$

$$\text{or} \qquad \sqrt{m} = \sqrt{m_0}\left(1 + \frac{Ve}{m_0c^2}\right)^{\frac{1}{2}}$$

is the nickel crystal and F is the Faraday ionisation chamber for receiving the electron beam after it is reflected from the crystal.

From the study of diffraction of X-rays by crystals we know that there is strong reflection when

$$2d \sin \theta = n\lambda \qquad \text{...(13)}$$

a result known as Bragg's law.

In this relation d, the spacing of successive crystal planes for a given crystal, is fixed and for a given setting of the experiment the *glancing angle* θ is also fixed. Now using electron beams of different velocities which can be done by gradually changing the voltages V of the electron gun, experiment showed several peaks in the Faraday chamber readings.

In one experiment from the known grating constant of nickel $d = 0.91 \times 10^{-8}$ cm ; glancing angle 0=65° and $n = 1$, Bragg's equation $2d \sin \theta = n\lambda$, gave $\lambda = 0.65 \times 10^{-8}$ cm. On the other hand, for 54 volt electron, the de Broglie wavelength given by the relation $\lambda = \frac{h}{mv}$ is computed to be 1.66×10^{-8} cm ; in excellent agreement with the experimental value. The wavelengths deduced from this diffraction experiment for other values of n also agreed very well with calculated values. These experiments provided a very clear evidence of an electron beam showing wave-like character.

These results were further corroborated by the experiments of Sir G.P. Thomson (Great Britain), who obtained transmission diffraction patterns by making the electrons strike thin crystal films.

G.P. Thomson's experiments. In 1928, G. P. Thomson employed high speed electrons ranging from 10^4 volts to 5×10^4 volts for his researches on electron waves. He shot mono-energetic beams of electrons through thin films of gold, platinum, aluminium etc. The thickness of the films used was of the order of 10^{-6} cm to obtain good results. In each case he could obtain a symmetrical pattern consisting of concentric rings with a central spot. These patterns were very much like those obtained in X-rays diffraction experiments. Since diffraction patterns can be produced only by waves, these experiments provided an excellent evidence for the wave-like behaviour of the electron beams. From the relative dimensions of the patterns obtained with X-rays and with electron beams the wavelengths associated with the electrons could be calculated. These values showed a good agreement with those calculated with the help of de Broglie's formula.

The schematic diagram of the apparatus employed by G.P. Thomson. Here F is a heated filament in the discharge tube AC in which cathode rays are produced with the help of a step-up transformer. The diaphragm A produces a fine pencil of electron beam which is allowed to fall on a thin film E. P is the photographic plate which receives the electrons after they are diffracted by the film.

Experiment with molecular beams. A few years later the German physicist O. Stern repeated the experiments of Davisson and Germer by using a molecular beam of sodium atoms and found that diffraction phenomenon exists in that case too. Dempster, Frisch, Eastermann and others succeeded in obtaining clear diffraction effects with hydrogen and helium by suitably reflecting them from crystal gratings thus confirming de Broglie's general concept of matter waves and the notion of 'pilot waves' guiding material particles in their motion.

A schematic sketch of the apparatus used by *O.* Stern. A beam of sodium atoms from an electrical oven *O* after passing through a diaphragm *D* is allowed to fall on a crystal C. The atoms scattered by the crystal in different directions are collected in small bottles B_1, B_2, B_3, etc. and their amounts are measured, the results show a strong maximum in the direction required by the ordinary law of reflection in accordance with Bragg's equation and number of secondary maxima were observed corresponding to diffraction pattern.

In Eastermann's experiment to study the wave behaviour of moving helium atoms, the gas was heated in an enclosure to 400°K from which a beam of atoms emerged through a slit. From kinetic theory of gases we can calculate the mean energy of atoms according to the equation

$$\frac{1}{2}\, Mc^2 = \frac{3}{2}kT \qquad \text{...(14)}$$

where k is Boltzmann's constant = 1.38×10^{-23} joule/°K.

Now the mass of the helium atom

$$MH_e = 6.7 \times 10^{-27} \text{ kg},$$

whence the root mean square velocity,

$$\bar{c} = \sqrt{\frac{3kT}{M_{H0}}} = \sqrt{\frac{3 \times 1.38 \times 10^{-23}\ \text{J}^{\circ}\ \text{K}^{-1} \times 400^{\circ}\ \text{K}}{6.7 \times 10^{-27}\ \text{kg}}}$$

$$= \sqrt{\frac{3 \times 1.38 \times 4}{6.7}} \times 102 \text{ m s}^{-1} \quad \text{...[J = kg-m}^2\text{s}^{-2}]$$

$$= 1.57 \times 10^2 \text{ m s}^{-1}.$$

The de Broglie wavelength of He atoms whose speed is 8.57×10^2 m s^{-1} is

$$\lambda = \frac{h}{M_{He}\, C}$$

$$= \frac{6.625 \times 10^{-24}\ J - s}{6.7 \times 10^{-27}\ \text{kg} \times 1.57 \times 10^{3}\ \text{m s}^{-1}}$$

$$= 6.\frac{6.625}{6.7 \times 1.57} \times 10^{-10}\ \text{m}$$

$$= 0.63 \times 10^{-10}\ \text{m}$$

$$= \mathbf{0.63\ Å.}$$

The emerging atoms had a Maxwellian distribution of velocities and a special arrangement was made to secure a beam of more uniform velocity. The beam was then made to strike the surface of a Li *F* crystal and scattered atoms in a given direction were collected in a small chamber as in Stern's arrangement. The diffraction patterns obtained corresponded very closely to the de Broglie wavelength calculated above.

The electron microscope. This is an interesting and highly useful application of the de Broglie matter waves. We have already seen that a beam of fast moving electrons behaves like extremely short waves. Such beams are made to replace the ultraviolet light in a microscope thus giving magnifications and revolving powers of a very high order. In the case of an electron moving in an evacuated tube under a potential difference of *V* volts, the associated de Broglie wavelength is given by

$$\lambda = \frac{12.27}{\sqrt{V}}\ Å \qquad \text{...}cf \text{ Eq. } 10$$

For $V = 10{,}000$ volts, the relation gives $\lambda = 0.122$ Å. This is much shorter than the wavelength of ultraviolet light, therefore, much greater resolving power could be obtained.

Let us now consider the design of an electron lens. The

basic idea is that of focusing a divergent electron beam by means of an electrostatic or a magnetic field in the same way as a beam of light is focused by a lens.

Example. *The de Broglie wavelength of an electron is 73 Å. What is its velocity ?* $m_e = 9.1 \times 10^{-31}$ *kg and* $h = 6.6 \times 10^{-24}$ *J-s*

$$\lambda = \frac{h}{mv}$$

$$v = \frac{h}{m\lambda}$$

$$= \frac{6.6 \times 10^{-34}\ \text{J} - \text{s}}{9.1 \times 10^{-31}\ \text{kg} \times 73 \times 10^{-10}\ \text{m}}$$

$$= \mathbf{9.93 \times 10^{4}\ m - s^{-1}}$$

Example. *Calculate the de Broglie wavelength in* A *associated with proton of kinetic energy 100 ev.* (*Mass of proton* = 1.6725×10^{-27} *kg and* $h = 6.625 \times 10^{-34}$ *J-s*).

The K.E. is given by

$$\frac{1}{2} mv^2 = 100 \text{ ev} = 100 \times 1.6 \times 10^{-19} \text{ J}$$

$$\therefore \quad v = \sqrt{\frac{2 \times 100 \times 1.6 \times 10^{10}\ \text{J}}{\text{m}}}$$

and momentum,

$$mv = \sqrt{200 \times 1.6 \times 10 - 19\ \text{J} \times \text{m}}$$

$$= \sqrt{3.2 \times 10^{-17}\ \text{J}\ 1.6725 \times 10^{-27}\ \text{kg}}$$

$$= 10^{-25} \sqrt{320 \times 16725\ kg\ m - s^{-1}}$$

Now $\qquad \lambda = \dfrac{h}{mv}$

Substituting the values of h and mv, we have

$$\lambda = \frac{6.625\times 10^{-34}\ J-s}{10^{-25}\sqrt{320\times 16725\ kg\ m-s^{-1}}}$$

$$= \frac{6625\times 10^{-37}}{10^{-25}\sqrt{320\times 16725}}\ \text{m}$$

$$= \frac{6625\times 10^{-12}}{2313}$$

$$= 2.864\times 10^{-10}\ \text{m.}$$

$$= \mathbf{0.02864\ \text{Å}.}$$

Example. *Find the de Broglie wavelength of a 10 kev electron, using SI units.*

The kinetic energy of the electron

$$K = \frac{1}{2}mv^2 = Ve$$

$$\therefore \qquad m^2v^3 = 2m\ Ve$$

or momentum

$$mv = \sqrt{2\text{m Ve}} = \sqrt{2\text{m}\times 10^4\ \text{ev}}$$

Substituting the values of m and ev, we get

$$mv = \sqrt{2\times 9.11\times 10^{-31}\ \text{kg}\times 10^4\times 1.6\times 10^{-19}\ \text{J}}$$

$$= 10^{-22}\sqrt{2\times 9.11\times 1.6\ kg-m\,s^{-1}}$$

$\therefore$ the de Broglie wavelength,

$$\lambda = \frac{h}{mv} = \frac{6.625\times 10^{-34}\ J-s}{10^{-23}\sqrt{2\times 9.11\times 1.6\ kg-m\ s^{-1}}}$$

$$= 1.227\times 10^{-11}\ \text{m} = 0.1227\times 10^{-10}\ \text{m}$$

$$= \mathbf{0.1227\ \text{Å}.}$$

Example. *Find the de Broglie wavelength of a 15 kev proton. Mass of proton (M_p) = 1.67 × 10^{-27} kg. Use SI units.*

The kinetic energy of the proton

$$\frac{1}{2} M_p v^2 = 15,000 \text{ ev}$$

$$= 15,000 \times 1.6 \times 10^{-19} \text{ J} = 24 \times 10^{-16} \text{ J}$$

$$\therefore (M_p v)^2 = 2 \times 24 \times 10^{-16} \text{ J } M_p$$

or momentum,

$$M_p v = \sqrt{48 \times 10^{-16} J M_p}$$

∴ the de Broglie wavelength,

$$\lambda = \frac{h}{M_p v}, \text{ where } M_p = 1.67 \times 10^{-27} \text{ kg}$$

$$= \frac{6.625 \times 10^{-34} J - s}{\sqrt{48 \times 10^{-16} J \times 1.67 \times 10^{-27} kg}}$$

Simplifying, we get

$$\lambda = \frac{6.625 \times 10^{-34}}{10^{-21}\sqrt{3 \times 1.6 \times 1.67}}$$

$$= 2.341 \times 10^{-13} \text{ m} = \mathbf{0.0023 \text{ Å}}$$

Example. *Calculate on the wave-mechanic theory the wavelength associated with (a) an electron moving with speed of the electron in the K-ring of the hydrogen atom (b) a molecule of the hydrogen at 0°C (its velocity=1.84 × 10^3 m/sec. at 0°C). Use UI units.*

(*a*) The speed of the electron in the *K*-ring of hydrogen atom,

$$v = 2.18 \times 10^6 \text{ ms}^{-1}$$

$$m = 9.11 \times 10^{-21} \text{ kg}$$

$$\lambda = \frac{h}{mv}$$

$$= \frac{6.625 \times 10^{-34}\ J-s}{9.11 \times 10^{-31}\ kg \times 2.18 \times 10^{6}\ m\ s^{-1}}$$

$$= 3.33 \times 10^{-10}\ \text{m}$$

$$= \mathbf{3.33\ \text{Å}.}$$

(*b*) $\lambda = \frac{h}{mv}$

The mass of a hydrogen molecule,

$$M = 2 \times 1.67 \times 10^{-27}\ \text{kg}$$

and $v = 1.84 \times 10^{3}\ \text{m}\ \ \text{s}^{-1}$.

$$\therefore \quad \lambda = \frac{6.626 \times 10^{-34}\ J-s}{2 \times 1.67 \times 10^{-27}\ kg \times 1.84 \times 10^{3}\ ms^{-1}}$$

$$= \frac{6.625}{3.33 \times 1.84} \times 10^{-10}\ \text{m}$$

$$= 1.07 \times 10^{-10}\ \text{m} = \mathbf{1.07\ \text{Å}.}$$

Wave mechanics. The subject of wave mechanics which may be termed as *modern quantum mechanics* originated in the fundamental discoveries of the dual aspect of both radiation and matter—*particle aspect* and *wave aspect*, and acceptance of Einstein's principle of equivalence of *mass and energy*. It was developed in the year 1925-27 by Erwin Schroedinger, Werner Heisendberg, Dirac and others and it arose from the fact that in dealing with small particles such as electrons the usual classical concepts of position and momentum do not apply (Heisendberg's uncertainty principle). It is the name given to the new methods of interpreting physical phenomena in the microscopic world that may provide the fundamental principle of recontacting the apparently conflicting views. In fact it goes beyond this and makes the ordinary gross mechanics of Newton merely a special case of a more general theory. It covers a wider ground than the old quantum theory of Bohr and Sommerfeld. By early 1930's, the application of new quantum mechanics to problems involving nuclei, atoms, molecules and matter in the solid state made it possible to understand a vast body of other puzzling data and even led to predictions of remarkable accuracy.

M. de Broglie has summarised the conception of wave mechanics as follows :

> "The fundamental postulate of wave mechanics is that with every independent particle, whether a matter or of radiation, there is associated the propagation of a wave, the intensity of this wave representing at each point and at each instant the probability that the associated particle will reveal itself at this point and at this instant. In the words of Max Born "The new quantum mechanics assumes that all laws of physics are of statistical character. The fundamental quantity is a wave function which obeys laws similar to those of an acoustical or optical wave ; it is not, however, an observable quantity but determines indirectly the probability of observed process."

Heisenberg's 'Uncertainty Principle'. One of the most novel and not easily visualized concepts in Wave Mechanics concerns the 'Uncertainty' which is naturally inherent in the simultaneous measurements of certain pairs of dynamical variables. The wave theory of light describes the motion of photons statistically but fails to specify in detail, how any particular photon will move. Similarly, de Broglie's theory of matter waves describes only the *probable* location of particles somewhere in the wave packet but fails to specify precise orbits like those assumed by Bohr. These wave theories therefore seem to have uncertainties built into them. An interesting interpretation of this duality has been given in 1927 by Werner Heisenberg in his 'uncertainty principle'. According to this, uncertainty with regard to the position of an electron within the wave packet and its probable velocity at any particular instant is an inherent property of nature, that is, we cannot know *in principle* the absolute precise values that will completely describe a physical system. In all physical measurements, *when one quantity is measured with more precision another related quantity is known with less precision,* however perfect the apparatus with which we are equipped and how great the skill to use it. Heisenberg's was no concern for experimental difficulties but it was the theoretical difficulty of measurement which he predicted. Referring to the simultaneous determination of position and momentum of the particle, the principle states that the uncertainty Δx involved in the measurement of coordinate x of the particle and the uncertainty Δp involved in the measurement of its momentum p are related by the equation

$$\Delta x . \Delta p \geq h, \quad \text{...(15)}$$

where h is Planck's constant. This means that if one variable is known more accurately, that is, smaller we make Δx, the larger becomes the conjugate variable Δp and *vice versa* so that the uncertainty principle holds. This principle is an expression of the wave particle duality of all physical entities and shows that it is impossible to measure both the position and momentum accurately simultaneously. As said above it does not depend upon the inaccuracies involved in the design and use of measuring instruments. This is one of the basic laws of the physical world since it sets a limit to the possible accuracy of observation and is of great philosophical importance.

The physical implications of Heisenberg's principle are well illustrated by analysis of certain 'idealized experiments'.

Position-momentum uncertainty

(*a*) *Determination of the position of a particle by a gamma ray microscope.* This microscope serves as an illuminating example of the principle and shows how the wave concept acts as limitation on the particle concept. A beam of mono-energetic electrons can be obtained experimentally. Let us assume that their velocity is known exactly and let a high resolving power microscope be used to obtain information on the position of an electron in the beam. We have already seen from classical wave theory that in the case of a microscope the minimum resolvable distance Δx is given by

$$\Delta x = \frac{\lambda'}{2 \sin \alpha},$$

where λ' is the wavelength of the light used, slightly altered as a consequence of impact with the electron and α is the semi-vertical angle of the cone of light coming from the illuminated object. Δx therefore represents the uncertainty in the determination of the x-coordinate of the position of the electron. For the electron to be observed at least one photon of energy $h\nu$ must be scattered into the microscope and this photon interacting with the electron, causes it to recoil. In the scattering process momentum is transferred from the photon to the electron (Compton effect) in an indeterminate direction. The scattered photon can enter the microscope anywhere within the semi-vertical angle a on either side. If it enters the right extreme of the cone the resolved component of its altered momentum $p\left(=\frac{h}{\lambda'}\right)$ along the x-axis is $\frac{h}{\lambda'}\sin\alpha$. The momentum imparted to the electron in x-direction will therefore be equal to

$$\frac{h}{\lambda} - \frac{h}{\lambda'} \sin \alpha,$$

where h/λ is the momentum of the incident photon. If, however, it enters the left extreme of the cone the momentum in the x-direction imparted to the electron is given by

$$\frac{h}{\lambda} - \frac{h}{\lambda} - \left(-\frac{h}{\lambda'} \sin \alpha\right) = \frac{h}{\lambda} + \frac{h}{\lambda'} \sin \alpha.$$

Thus the momentum given to the electron can have any value between those given by these expressions. The uncertainty in the value of the momentum of the electron is therefore

$$\Delta p = \left(\frac{h}{\lambda} + \frac{h}{\lambda'} \sin \alpha\right) - \left(\frac{h}{\lambda} - \frac{h}{\lambda'} \sin \alpha\right)$$

$$= \frac{2h}{\lambda'} \sin \alpha .$$

The product of the uncertainties in the determination of the *simultaneous* values of the position and the momentum of the electron is, therefore,

$$\Delta x.\ \Delta p = \frac{\lambda'}{2 \sin \alpha} \times \frac{2h}{\lambda'} \sin \alpha = h$$

as predicted by Heisenberg's principle.

The error in any actual experiment will be greater than h which is the case in this ideal experiment. Hence

$$\Delta x\ \Delta p \geq h$$

(*b*) *Diffraction of a beam of electrons through a narrow slit.* Let us next consider the uncertainties involved in the *simultaneous* determination of position and momentum of an electron when an electronic beam passes through a narrow slit *AB* and the diffraction pattern is recorded on a photographic plate.

If Δy is the width of the slit, then the uncertainty in the y-coordinate perpendicular to the direction of flight of the electron is Δy, as it is indefinite at what place it has passed through the slit. The narrower the slit, the less the indeterminance in knowledge of the coordinate y.

But a very narrow slit produces a definite diffraction pattern on the photographic plate. This diffraction pattern is due to the fact that the electron receives additional momentum parallel to the slit width at the instant it passes through it. If p is the momentum of the electron and α the angle of deviation, the component of momentum in the y-direction is $p \sin \alpha$. The diffracted electron may be anywhere within the observed pattern, so that the uncertainty in the value of the y-component of the momentum of the electron is

$$\Delta p = p \sin \alpha.$$

The mean angle of deviation, *viz.*, angular width of the diffraction pattern is determined by the slit width Δy and is given by the equation

$$\Delta y = \frac{\lambda}{\sin \alpha}$$

Hence the product of the uncertainties is given by

$$\Delta y \cdot \Delta p = p \sin \alpha . \frac{\lambda'}{\sin \alpha} = p\lambda.$$

From de Broglie's hypothesis $p = h/\lambda$

$$\therefore \qquad \Delta y . \Delta p = p\lambda = \frac{h}{\lambda'} \times \lambda = h.$$

In an actual case again as before $\Delta y . \Delta p \geq ph$

Energy-time uncertainty. Another set of variables can be used to express Heisenberg's uncertainty principle. If E is the energy of a physical system at time t, then it can be shown that

$$\Delta E . \Delta t \geq h, \qquad ...(15\ a)$$

where ΔE is the uncertainty in the knowledge of the value of E and Δt is the uncertainty in the knowledge of the value of time t taken to measure it. In other words, *the uncertainty of an energy times the time spent in its measurement is at least of the order of magnitude of Planck's constant. This is Einstein's version of the uncertainty principle.* It shows that the uncertainty of an energy measurement depends upon the time available to make that measurement. Thus if an atom remains in the excited state for a *very* short time, the precise energy of the state may not be determined very accurately.

According to this version the relation emerges from the following consideration :

The velocity v of a wave packet can be determined by observing the position of the particle at two different times. If the packet has width Δx, then the momentum of the particle is uncertain within Δp, as given by equation (15) and the energy is indeterminate within

$$\Delta E^* = \frac{p}{m} \Delta p = v \Delta p.$$

Now the uncertainty in time is of the order $\Delta x/v = \Delta t$, whence

$$\Delta E. \Delta T = v \Delta p \frac{\Delta x}{v} = \Delta x \Delta p = h.$$

...[*cf.* equation (15)

or in an actual experiment

$$\Delta E. \Delta T \geq h$$

Thus once again there is no escape from uncertainty relation. The conclusion forces us to abandon the law of exact causality in the classical sense. In fact probability takes the place of exactness in physical science. The phenomena are therefore to be redescribed in terms of probability and statistical distribution.

The uncertainty principle may be regarded as one of the profound truths of nature and it is applicable even beyond the field of physics, where it originated.

A more realistic method of approach yields for the uncertainty principle

$$\Delta x.\ \Delta px = \frac{h}{2\pi} \qquad ...(16)$$

an expression more widely used than Eq. (15).

Application. An important application of the principle lies in the determination of the absence of an electron in the nucleus of an atom thus confirming the untenability of the proton-electron hypothesis of nucleus. As a typical nucleus is less than 10^{-14} m in radius, it is obvious that in order that an electron may lie within such a nucleus, the uncertainty in its position may not exceed 10^{-14}m. The corresponding uncertainty in the momentum of the electron is

$$\Delta p \geq \frac{h}{2\pi\Delta x} \leq \frac{6.625 \times 10^{-34}\ J-s}{2\pi \times 10^{-14}\ m}$$

$$\geq 1.1 \times 10^{-20} \text{ kg m s}^{-1}$$

The momentum of the electron therefore must be at least comparable in value with this magnitude. Now an electron having momentum p equal to 1.1×10^{-20} kg ms^{-1} has a kinetic energy, K many times greater than its rest mass energy m_0c^2 and we may accordingly use the extreme relativistic formula

$K = pc$ to find the kinetic energy.

Substituting for p and c, we obtain

$$K = 1.1 \times 10^{-20} \text{ kg ms}^{-1} \times 3 \times 10^{8} \text{ ms}^{-1}$$

$$= 3.3 \times 10^{-12} \text{ J}$$

Since Mev = 1.6×10^{-13} J, the kinetic energy of the electron must be well over 20 Mev, if it is to be nuclear constituent. Experiments show that electrons associated even with unstable atoms never have more than a fraction of this energy and hence we conclude that electrons cannot be present within nuclei of atoms.

Schroedinger's wave equation. If the particle has wave properties as de Broglie proposed it is expected that there should be some sort of wave equation which describes the behaviour of the *wave function* (the variable quantity characterising the de Broglie Waves). Accordingly by analogy with a typical wave equation Schroedinger derived a mathematical relation to provide a statistical approach to wave particle dualism. The main problem envisaged by him was the calculation of the *probability* of finding the electron at various points in its orbit. For material particles the necessary wave functions were obtained from the solution of this wave equation first formulated by him and called *Schroedinger's Wave Equation.* It works for atomic physics and forms the basis of this new division of physics called wave mechanics.

In one form characterising the steady state (time independent) it is most briefly represented in three dimensions as under :

$$\frac{\partial^2 \psi}{\partial x^2} + \frac{\partial^2 \psi}{\partial y^2} + \frac{\partial^2 \psi}{\partial z^2} + \frac{8\pi^2 m}{\partial h^2} (E - U) \psi = 0 \quad ...(17)$$

or

$$\nabla^2 \psi + \frac{8\pi^2 m}{h^2} - (E - U) \psi = 0 \quad ...(18)$$

where ψ (pronounced as S_i) is usually referred to as the *wave function, E* denotes the total energy of the particle and *U* its potential energy which varies with the position of the particle only, *m* the mass of the particle, *h* is the Planck constant and

$$\nabla^2 \psi \equiv \frac{\partial^2}{\partial x^2} + \frac{\partial^2}{\partial y^2} + \frac{\partial^2}{\partial z^2}$$

read as 'del squared' denotes the usual Laplacian operator in vector notation. The wave function ψ is the variable quantity which characterises the de Broglie waves associated with matter in the same manner as the variation in the magnetic field in space and time constitute the light waves and those of pressure in space and time constitute the sound waves. The value of the wave function associated with a moving body at the particular point x, y, z in space at the time t is related to the probability of finding the body there at the time. The problem in wave mechanics is to insert the appropriate values of U and seek solution of Schroedinger's equation. This solution will give ψ as a finite single-valued continuous function of the coordinates in space. This function is synonymous with the *amplitude* of the wave. But the amplitude of any wave may be both *positive* and *negative* and since a negative probability has no meaning, the value of the square of the wave function $|\psi^2|$ is taken as a measure of probability density.

The probability therefore of finding the particle at any given point in the space at a given time t is $|\psi^2|$, which is independent of t. The larger the value $|\psi^2|$, the stronger is the possibility of the body's presence at a specified position and *vice versa*.

These concepts are in complete agreement with classical mechanics and wave mechanics could therefore be used to solve problems in mechanics that occur in engineering practice. For instance, the application of this equation to the electronic conduction of metals is essentially of finding the ϕ values for the electrons in the metals and from these determining the energies of the electrons. The application of the Schroedinger equation to the determination of the energies of the electrons associated with individual atoms yields the same results that had to be postulated in the original Bohr and Sommerfeld theories of the atom. For example, substituting for the potential energy U the term $\frac{Ze^2}{4\pi \epsilon_0 r}$ (SI system) for hydrogen like atoms the Schroedinger equation becomes

$$\nabla^2 \psi + \frac{8\pi^2 m}{h^2}\left(E + \frac{Ze^2}{4\pi \epsilon_0 r}\right) \psi = 0 \qquad ...(19)$$

This equation could be solved in terms of known functions and the solution existed and represented standing waves only for discrete values of E. The solution for E yields

$$E_n = \frac{Z^2 me^4}{8 \epsilon_0^2 h^2 n^2} \qquad ...(20)$$

which is identical to the Bohr equation (12), obtained on the basis of classical mechanics introducing arbitrarily the postulate that angular momentum is quantized. In this equation the term n corresponds to the principal quantum number considered earlier.

Obviously, therefore, the Bohr model may be considered as a direct consequence of wave-mechanical approach. Similarly, if one put in for U the value of potential energy for a linear oscillator, he got the correct energy levels for this, and so on for other problems.

In other words, here was a mathematical foundation which led to the same results as Bohr's theory but in a much more reasonable and fundamental way so that more logical picture of hydrogen atom emerged. At the same time the various problems raised by the Bohr theory of complex spectra and the intensities of their spectral lines were automatically resolved. For instance, in hydrogen the quantum number l equal to zero for an s state, 1 for a p state and so on automatically entered in the correct way, and the other puzzles of the older theory removed. Schroedinger showed how the equation could be formulated for any mechanical problems whatever, so that it was not limited in any way to the multiplying periodic problems to which alone Bohr theory could be applied. He showed how to set it up for many-electron problem. The mathematics became extremely complicated, so hard that scientists are still working on methods of solving it; but there was no doubt that the formulation was right.

Wave-mechanical concept of the atom

(*i*) *De Broglie's quantization of orbits.* The wave-mechanical treatment of an atom illuminates many of the old *ad hoc* assumptions made in the previous sections. For instance, on wave-mechanical model, a moving electron is associated with a wavelength X= h/mv. [*cf.* equation (2)] where m is the mass and v the velocity of the electron and h is Planck's constant. Instead of assuming the eléctron to be moving in a fixed stationary orbit it was imagined to generate a series of waves as it revolved around the nucleus in any permissible orbit. In order that waves should not cancel each other by interference in any permissible orbit the electron must move in such a manner as to produce a *standing wave* in the orbit. This will happen only if the electron wave just fills the circumference, where there are four complete wavelengths contained in it. The circumference $2\pi r$ must contain an integral number of wavelengths, stated mathematically,

$$2\pi r = n\lambda, \qquad ...(21)$$

where r is the radius of the orbit and n an integer.

Substituting the value of λ $(=h/mv)$ from de Broglie equation,

we get $$2\pi r = \frac{nh}{mv}$$

or $$mvr = \frac{nh}{2\pi} \qquad ...(22)$$

But mvr is the angular momentum of the electron regarded as a particle. Thus we see that the wave mechanical picture leads to Bohr's postulate No. (1) regarding the quantization of orbits—that the only permitted orbits are those whose angular momentum is equal to an integral multiple of $h/2\pi$.

This indicates that the electron does not move in any defined path but vibrates in a *haphazard* manner and at the same time is constrainted to be in the region of highest probability for most of the time. *This region in space where there is the greatest possibility of finding the electron is called an* orbital. Thus an orbital is the spatial description of the motion of an electron corresponding to a particular energy level. The energy of the electron in an orbital is always the same. The probability plot against the distance r from the nucleus for quantum numbers $n = 1, l = 0$; $n = 2, 1 = 1$; $n = 3, l = 2$. These quantum numbers correspond to those representing the circular orbits in Bohr's theory. It is seen that the probability of finding the electron at any distance is the greatest at the distance equal to the radius of the corresponding Bohr orbit.

This wave concept of the electron rather than as a particle revolving in a fixed obit justifies Bohr's second arbitrary assumption that an atom in a stationary state will not radiate energy because the radiations from the different parts of the electronic cloud will cancel each other by interference.

Further, in the wave mechanics the emission of photon involves an energy change of the atom in the two states ; mechanism of emission does not have to be introduced as a separate postulate as in Bohr's theory. This leads to the frequency condition.

$$\nu = \frac{E_i - E_f}{h} \qquad ...(23)$$

The initial and final states of a hydrogen atom that corresponds to quantum numbers n_i and n_f have from equation (*iv*) as :

$$\text{Initial energy, } E_i = \frac{me^4}{8 \epsilon_o^2 h^2}\left(\frac{1}{n_i}\right)^2$$

$$\text{Final energy } E_f = \frac{me^4}{8 \epsilon_o^2 h^2}\left(\frac{1}{n_f}\right)^2$$

The energy difference between these two states is

$$E_i - E_f = \frac{me^4}{8 \in_o^2 h^2}\left[\left(-\frac{1}{n_i}\right)^2 - \left(-\frac{1}{n_f}\right)^2\right]$$

$$= \frac{me^4}{8 \in_o^2 h^3} - \left[\frac{1}{n_f^2} - \frac{1}{n_i^2}\right]$$

The frequency of the photon released in the transition is

$$v = \frac{E_i - E_t}{h}$$

$$= \frac{me^4}{8 \in_v^2 h^3} - \left(\frac{1}{n_f^2} - \frac{1}{n_i^2}\right) \qquad ...(24)$$

Eq. 24, derived from Wave Mechanics agrees with Bohr's Eq. (23-8) deduced considering the electron as a particle circulating round the nucleus in a stationary orbit coupled with two arbitrary postulates.

Shortcomings of Bohr's theory removed. It can be easily seen that three main shortcomings inherent in Bohr's postulates are all satisfactorily removed by Wave Mechanics, namely—

(*i*) Bohr's postulate that the angular momentum of the revolving electron is equal to an integral multiple of $h/2\pi$ comes out as a natural consequence of the concept of the election producing standing waves in the orbit since in a standing wave transfer of energy is zero.

(*ii*) The same concept of the electron rather than as a particle revolving in an orbit also removes the second shortcoming in Bohr's theory that the electron travels in its orbit without radiating.

(*iii*) It can also be shown that in Wave Mechanics the emission of photon of frequency ν involves an energy change of the atom, the mechanism of emission does not have to be introduced as a *separate* postulate as in Bohr's theory.

Example. *An electron initially at rest is accelerated through a p through a p. d. of 5000 V. Compute (i) the momentum (ii) the de Broglie wavelength and (iii) the wave number of the electron. Also calculate the Bragg angle for its first order reflection from the (1,1,1) planes of nickel which are 2.04 Å apart.*

(*i*) Let v be the velocity of the lectron then its K.E.

$$\frac{1}{2}mv^2 = Ve = 5000 \text{ ev}$$

$$= 5000 \times 1.6 \times 10^{-19} \text{ J}$$

$$\therefore \quad (mv)^2 = 2\,m \times 5000 \times 1.6 \times 10^{-19} \text{ J}$$

and momentum of the electron,

$$mv = \sqrt{2\text{m} \times 5000 \times 1.6 \times 10^{-9}\, J}$$

$$= \sqrt{2 \times 9.11 \times 10^{31} \text{ kg} \times 5000 \times 1.6 \times 10^{-19}\, J}$$

$$\ldots [\ \therefore m = 9.11 \times 10^{-31} \text{ kg}$$

$$= \sqrt{9.11 \times 1.6 \times 10^{-46}} \text{ kg m s}^{-1}$$

$$= 10^{-23}\sqrt{9.11 \times 1.6} \text{ kg m s}^{-1}$$

$$= \mathbf{3.818 \times 10^{-28} \text{ kg ms}^{-1}}$$

(*ii*) The de Broglie wavelength of the electron

$$\lambda = \frac{h}{mv} = \frac{6.625 \times 10^{-34} \text{ J s}}{3.818 \times 10^{23} \text{ kg m s}^{-1}}$$

$$= 1.729 \times 10^{-11} \text{ m}$$

$$= \mathbf{0.1729} \text{ Å}.$$

(*iii*) Here wave number is the angular wave number given by

$$k = \frac{2\pi}{\lambda} = \frac{2\pi}{1.729 \times 10^{-11}\ m}$$
$$= 36.34 \times 10^{10}\ \text{m}^{-1}.$$

(*iv*) Let θ be the Bragg angle for the 1st order reflection from (1, 1, 1) planes of nickel (simple cubic lattice).

Here $d = 2.04$ Å ;

$n = 1$

and $\lambda = 0.1729$ Å.

Now by Bragg's equation

$$2\ d \sin\theta = n\lambda.$$

$\therefore 2 \times 2.04$ A $\sin\theta = 1 \times 0.1729$Å

or $$\sin\theta = \frac{0.1729}{2 \times 2.04} = \frac{0.1729}{4.08} = 0.04237$$

$$\theta = 2°, 25'.$$

3

Wave Packets

In discussing the phenomena of light and electromagnetic waves we dealt with harmonic waves of a single frequency but sometimes two or more harmonic waves of nearly equal frequencies superpose so that the resultant wave has a beat like modulate character as in sound. Now we shall study how such a wave propagates and carries energy and signals. We come across such a wave, when we throw pebbles in a water pond. We see that waves thus generated proceed outwards from the centre of origin in the form of circular arcs and soon convert into a wave packet, which carries energy with it and the pieces of cork lying at some distance from the centre of origin are disturbed and begin to jostle up and down as wave packet arrives.

On looking closely, we will see that the primary wavelets which make up the wave-packet do not maintain a constant position relative to the wave packet. They appear to vanish at the head and new wavelets emerge from the rear. The wave packet travels with a modulated or the so called group velocity different from that of the primary wavelets.

Wave Velocity and Group Velocity

We have seen that an individual sine wave propagates through a medium with a definite velocity and hence the phases (crests and troughs or compressions and rarefactions) of the wave advance with the same velocity.

The velocity of propagation of the phases is called wave or phase velocity. But when a disturbance consists of a number of component waves each travelling with slightly different velocity, the resultant (measured) velocity will be that of a periodicity impressed upon the wavetrain. The velocity of advance of this periodicity is called the group velocity.

Consider a wave train having two wavelengths in which the dotted curve represents the longer wave travelling faster than the shorter. At the instant considered the two are in phase at the point *A* giving rise to a maximum amplitude. Some time later the amplitude will build up a little to the left of the point *A*. In other words, the group will advance less rapidly than the separate trains. The energy belongs to the group rather than to the waves and the observed velocity will be that of the group.

In order to obtain an analytical relationship between the wave velocity and the group velocity, consider two infinite trains of waves of equal amplitude represented by

$$y_1 = a \sin (\omega t + kx) \qquad \text{... } (i)$$

and

$$y_2 = a \sin [(\omega + d\omega)\, t + (k + dk)x] \qquad \text{...}(ii)$$

where ω, $\omega + d\omega$ are their angular frequencies, be superimposed.

The resultant displacement at any time t and at any position $\underline{x}$ is given by

$$y = y_1 + y_2$$

$$= a \sin (\omega t + kx) + a \sin [(\omega + d\omega)\, t + (k + dk)x]$$

$$= 2a \sin \frac{1}{2} [(2\omega + d\omega)\, t + (2k + dk)\, x]$$

$$\times \cos \frac{1}{2} (d\omega t - dkx) \qquad ...(1)$$

Since $d\omega$ and dk are small as compared to ω and k respectively, we can write

$$2\omega + d\omega = 2\omega$$

and $$2k + dk = 2k$$

$$\therefore \quad y = 2a \left[\sin(\omega t + kx) \cos\left(\frac{d\omega t}{2} - \frac{dkx}{2} \right) \right] \qquad ...(1a)$$

Thus the resultant consists of a vibration of angular frequency ω and propagation constant k superimposed by a periodicity of angular frequency $d\omega/2$ and propagation constant $dk/2$. The effect of periodicity is to produce successive wave groups.

The wave or phase velocity is given by

$$v_\phi = \omega lk \qquad ...(2)$$

while the velocity of the wave group,

$$\therefore \quad v_g = d\omega/dk \qquad ...(3)$$

since $\omega = kv_\phi$ ... by Eq. (2).

$$\therefore \quad v_g = d\omega/dk = v_\phi + \frac{dv_\phi}{dk} \cdot k$$

since $$k = \frac{2\pi}{\lambda},$$

$$\therefore \quad \frac{dk}{d\lambda} = -\frac{2\pi}{\lambda^2}$$

or $$dk = \frac{2\pi}{\lambda^2}\, d\lambda$$

$$\therefore \qquad v_g = v_\phi + \frac{dv_\phi}{-2\pi\, d\lambda / \lambda^2} \cdot \frac{2\pi}{\lambda}$$

or
$$v_g = v_\phi - \lambda \frac{dv_\phi}{d\lambda} \qquad ...(4)$$

If $v_\phi = c$, the velocity of light

$$v_g = c - \lambda \frac{dc}{d\lambda}.$$

The distinction between wave velocity and group velocity can be observed when a train of wave of approximately the same wavelength is advancing over the surface of deep water. The individual waves are seen to advance through the group, dying away as they approach its anterior limit, while their places are taken by the fresh waves which appear at the rear. The group advances as well but not as rapidly as the waves.

The relation between v_ϕ and v_g in the case of water waves is discussed below.

Water Waves

When water surface in a tank, say, is disturbed, there are two main kinds of forces, besides hydrostatic pressure and viscosity, that tend to restore the surface to the normal form, one is gravity when the waves are large but for smaller waves it is the surface tension that predominates. The particles of the liquid under the joint action of surface tension and gravity execute a periodic elliptic motion. It can be shown that the velocity of propagation of surface wave is given by

$$v = \sqrt{\frac{g\lambda}{2\pi} + \frac{2\pi T}{\lambda\rho}} \qquad ...(5)$$

where ρ is the density of water= 1 gm/cc, g the acceleration of gravity = 981 cm/sec^2 and T *is* the surface tension of water = 73 dynes per cm.

Waves and Ripples

When λ is large, obviously only the first term under the radical is important, whereas if λ is small only the second term becomes effective. For wavelengths in water longer than 10 cm, the surface tension term may be neglected and the velocity is then given by

$$v = \sqrt{\frac{g\lambda}{2\pi}}$$

These are known as gravity waves on the surface of deep water, since gravity predominates, while for smaller wavelengths, less than 3.0 mm, the term $g\lambda/2\pi$ becomes negligible so that the velocity of the wave $v = \sqrt{\frac{2\pi T}{\lambda\rho}}$. These waves are called *ripples* and their propagation is mainly due to surface tension. If λ lies in between these two values, both terms should be included in the determination of velocity. Considering the variations of velocity of the phase wave it tends to infinity both for very small and very large values of λ, we therefore note that there must be a wavelength of intermediate value *(critical wavelength)* for which velocity of water waves is *minimum*. We can show that this occurs at a wavelength of 1.70 cm, when the minimum velocity (v_{min}) is 23 cm per sec. The phase and group velocities are then both 23 cm/sec. For wavelengths much less than 1.70 cm, the group velocity is 1.5 times the phase velocity. For wavelengths much greater than 1.70 cm the group velocity is half the phase velocity.

Example. *The phase velocity of ocean waves is* $\sqrt{g\lambda / 2\pi}$ *where g is the acceleration due to gravity. Show that group velocity of these waves is half the phase velocity.*

The group velocity of these waves by Eq. (4) is given by

$$v_g = v_\phi - \frac{\lambda dv_\phi}{d\lambda} \qquad ...(i)$$

v_ϕ is the phase velocity and λ is the wavelength of ocean waves.

Now $\quad v_\phi = \sqrt{g\lambda/2\pi} \qquad$ (given)

$$\therefore \quad \frac{dv_\phi}{d\lambda} = \sqrt{\frac{g}{2\pi}} \cdot \frac{1}{2} \lambda^{-1/2}$$

$$= \frac{1}{2}\sqrt{g/2\pi\lambda}\,.$$

Setting the values of v_ϕ and $dv_\phi/d\lambda$ in Eq. (*i*) above, we have

$$v_g = \sqrt{g\lambda/2\pi} - \times\frac{1}{2}\sqrt{g/2\pi\lambda}$$

$$= \sqrt{g\lambda/2\pi} - \times\frac{1}{2}\sqrt{g\lambda/2\pi}$$

$$= \frac{1}{2}\sqrt{g\lambda/2\pi} = \frac{1}{2}v_\phi$$

that is, group velocity is one half of the phase velocity.

Phase and group velocities for non-dispersive waves. In radio broadcasting the electromagnetic carrier wave consisting of, say, two components of frequencies ω_1 and ω_2 travel with an average frequency $\frac{\omega_1 + \omega_2}{2}$ of about 1000 kilocycles (*kc*) per sec, while the signal frequencies superimposed on it, will be audible frequencies in the range of 20 cps to 20 keps. This modulated carrier wave will be of the form of wave packets. This modulated carrier wave propagates not at average phase velocity $v_{av} = \omega lk$ but at the modulated or group velocity $v_a = d\omega/dk$. We give below some physical examples of phase and group velocities for non-dispersive waves.

1. *Electromagnetic waves in vacuum.* The dispersion relation for these waves is given by

$$\omega = ck.$$

The phase and group velocities are

$$v_\phi = \omega/k = c, \; v_g$$
$$= d\omega/dk = c \qquad ...(6)$$

Thus phase and group (modulation) velocities are both equal to c for electromagnetic waves in vacuum.

2. *Other non-dispersive waves.* Light waves in vacuum are non-dispersive, *i.e.*, the phase velocity does not depend upon frequency. In such a case, phase and group velocities also agree, since in general, we have

$$\omega = v_\phi k$$

and
$$v_g = \frac{d\omega}{dk}$$
$$= v_\phi - \frac{\lambda dv_\phi}{d\lambda}$$

Thus in inter-stutter space the group and phase velocities are equal as $dv_\phi/d\lambda=0$, v_ϕ being equal to c (a constant).

It is $v_\phi(=c)$ that is measured by Roemer's astronomical method of determining the velocity of light. The terrestrial methods give the group velocity v, $\left(= v_\phi - \lambda \frac{dv_\phi}{d\lambda}\right)$, but when the medium is air, the correction term $\lambda \frac{d\nu_\phi}{d\lambda}$ being small can be neglected. In Michelson's determination of the velocity of light in carbon disulphide (CS_2) by the rotating mirror method the correction term has to be taken into account to get the phase velocity v_ϕ *i.e.*, *c*. He found the velocity of light in air 1.76 times the velocity in carbon bisulphide whereas index of refraction method would give the velocity as 1.64 times. The difference arises because the index of refraction gives the ratio between the phase velocities $\left(\frac{v_\phi \; air}{v_\phi \; CS_2}\right)$ while experiment measured the group velocity in CS_2,. By applying the correction term the two results were found to agree.

Audible sound waves are another example of non-dispersive waves, where we have

$$\omega = v_\phi k$$

$$= \sqrt{\frac{\gamma p}{\rho}}\, k \qquad ...(7)$$

and similarly in the case of *transverse waves* in a continuous string.

$$\omega = v_\phi k$$

$$= \sqrt{\frac{T}{m}}\, k\,. \qquad ...(8)$$

4

Quantum Theory

The beginning of the twentieth century was the revival of Newton's corpuscular theory of light in a modified form. In 1900 Max. Planck, a German theoretical physicist, in order to explain the observed energy density in a radiation field put forward the hypothesis, of *atomicity* of energy and introduced his *quantum of action*. It has since then succeeded in explaining many different phenomena concerning the interaction between energy and matter. According to this theory an accelerated electron, which in the original hypothesis he described as a linear oscillator, does not radiate energy continuously as demanded by the electromagnetic theory of light but it does so in tiny *packet* or *quanta*. The energy in any one of these quanta is proportional to the frequency of the radiation emitted. Thus if ν is the frequency of the radiation, the energy content of each quantum is expressed by

$$E = h\nu \qquad ...(1)$$

where h is a universal constant known as Planck's constant. Its value is 6.625×10^{-34} J-s. These quanta are basic units of energy and cannot be subdivided. They are as it were atoms of energy and are not called photons.

For instance, for yellow light from a sodium flame of wavelength 5896 x 10^8 cm and hence of frequency

$$\nu = \frac{c}{\lambda} = \frac{3 \times 10^{10} \text{ cm s}^{-1}}{5896 \times 10^{-8} \text{ cm}} = 509 \times 10^{12} \text{ s}^{-1}$$

the energy of the yellow light photon is give by

$$E = h\nu$$

$$= 6.625 \times 10^{-34} \text{ J-s} \times 509 \times 10^{12} \text{ s}^{-1}$$

$$= 3.372 \times 10^{-19} \text{ J-s}$$

These photons travel in space much like Newton's corpuscles. They remain concentrated in a finite number of energy corpuscles moving like material particles and do not spread out evenly like electromagnetic waves.

In Planck's original form of the theory, only the emission was supposed to occur in quanta whereas absorption took place continuously until it reached the quantum value for the radiation concerned but this view has since been modified. We shall now discuss the photoelectric effect and other evidence in favour of quantum theory.

The photoelectric effect. When ultra-violet light and in some cases even visible light is allowed to fall on a metal plate it is found that electrons are emitted. This emission stops as the potential of the metal plate increases and resists the escape of electrons. The phenomenon is called the photoelectric effect and the electrons emitted are known as *photoelectrons* as they are emitted by the action of light. The effect was first observed in 1887 by Heinrich Hertz who found that an electric spark started more readily when the negative electrode was exposed to ultra-violet light and was discovered a year later (1888) by Hallwachs, who exposed a freshly polished zinc plate to the action of ultra-violet light from an arc lamp and found that it acquired a positive potential of 1 volt.

As regards the action of ultra-violet light the metals may be arranged in the order of increasing sensitivity, *viz, Zn, Mg, Li, Na, K, Kb, Ce* etc. In the case of potassium and even more so with rubidium and caesium, the emission of electrons takes place under the action of visible light.

The emission of electrons takes place immediately the light reaches the surface and ceases as soon as it is cut off. If the body be negatively charged it is discharged by the emission of electrons and if it is neutral, the photoelectric emission leaves it with *positive* charge. For a given intensity of the incident beam, the rate of emission depends upon the *pressure*, the *nature* of the gas around the plate and the *state of polish* of the surface. The effect is shown by non-metals, liquids and gases as well. In the case of non-metals owing to their more electro-negative character, the maximum wavelength at which the effect commences is smaller than in the case of metals.

Study of photoelectric effect. The effect can be easily studied with the apparatus.

A is a highly evacuated quartz-tube with two plates *P* and *Q* serving as electrodes to which a potential difference *V* can be applied by a battery *E*. When the plate *P* is exposed to the action of light, a current flows only if *Q* is *positive* with respect to P. This indicates that negatively charged particles are liberated by the action of light.

If the applied potential difference is increased, the current also increases till it reaches a limiting value called the *saturation current*. If, however Q is made negative with respect to *P*, the current diminishes till at a certain fixed negative potential difference the current goes down to zero. This value of potential difference is called the stopping potential. Its magnitude enables us to calculate the maximum kinetic energy of the electrons liberated from the plate P. If V is the stopping potential e, m, and v, the charge, mass and maximum velocity respectively of the electrons liberated, then

$$\frac{1}{2} mv^2 = eV,$$

whence
$$v = \sqrt{\frac{2eV}{m}}. \qquad ...(2)$$

Actually, however, the electrons emitted have various velocities upto a certain maximum but as those emitted below the surface lose a part of their energy in passing through the material it is reasonable to suppose that the velocity with which the electrons are liberated from the atoms is the same for them all and equal to the maximum velocity v.

For the same metal and the same frequency of light, the stopping potential is found to be independent of the intensity of light. If now the frequency (ν) of the incident light is gradually changed and the corresponding stopping potentials measured, a linear relation is found to hold between them.

It is further found that for each substance there is a limiting frequency of radiation that can cause emission. For all frequencies below that value, the emission does not take place, whereas all frequencies above that limit cause emission, This limiting frequency is called photoelectric threshold and depends upon the *nature* of the metal. The emission of electrons is governed by the following two experimental laws :

Law I. The number of electrons emitted by a given surface under the action of a beam of light is directly proportional to the intensity of the beam.

Law II. The velocity and hence the energy of the electrons leaving a metal under the influence of light is independent of the intensity of light and depends only on the frequency of the incident light and the kind of metal.

Explanation of photoelectric effect. The electromagnetic theory of radiation which seemed to be firmly established by the end of the last century was at first applied to explain the phenomenon, as it was supposed that the electric and magnetic fields which were attributed to be associated with the light waves may well exert forces on the electrons contained in the metal and so liberate them from the metal surface. On this theory we should predict that light of high intensity would give high kinetic energy to the electrons liberated. Furthermore, according to this theory we should expect that change in the frequency of radiation should produce little effect on the emitted electrons and a light of low frequency if sufficiently intense would be as effective as high frequency light. But the experimental facts stated above are quite otherwise.

Are we then to aspect that the *threshold* frequency is the resonance effect ? But this assumption too, is ruled out as according to resonance phenomenon the emission should fall off for frequencies above as well as below the threshold frequency whereas the experiment shows that emission continues with increased velocity when the frequency of the incident light is greater than the threshold frequency.

Evidently, therefore, the electromagnetic theory was unable to account for the phenomenon so Planck's quantum theory was applied by Einstein (1905) to explain the photoelectric emission.

According to this theory as has already been explained, whenever radiation was emitted or absorbed by matter, the exchange must take place in definite packets of energy $h\nu$ at a time, where h is Planck's constant and ν the frequency of incident radiation. Einstein extended this idea and suggested that in inducing the emission of photo-electrons light too does act like a stream of discrete quanta or photons. Each photon of energy $h\nu$ is so concentrated that when it is incident on a metal surface, it can *transfer* its whole energy content to a single electron. Conservation of energy then leads to Einstein's photoelectric law

$$h\nu_0 = \omega \qquad ...(3)$$

Here, E is the kinetic energy with which the photo-electrons leave the metal surface and ω is the *work function,* which depends upon the nature of the metal. Of the energy $h\nu$ of the incident photon a portion ω is required to extract the electron from the nucleus of the atom and to get out of the surface of the metal. The remainder E is imparted to the photo-electron as kinetic energy. So if $h\nu$ is less than ω there is no emission, however great the intensity of the incident beam, for an increase in intensity merely increases the number of photons per sq cm/sec, but it does not affect the energy content in each photon. Hence an electron can never get enough energy for its liberation from the metal surface. So *threshold effect* follows at once on this theory, the threshold frequency ν_0 being given by

$$h\nu = \omega, \qquad ...(4)$$

where ω, the work function, is different for different metals. If the energy $h\nu$ of the incident photon is greater than ω $(= h\nu_0)$ the electrons will be emitted with a maximum velocity v given by

$$\frac{1}{2}mv^2 = h\nu - h\nu_0.$$

$$= h(\nu - \nu_0). \qquad ...(5)$$

The quantum theory, therefore, suggests that *firstly* energy of the emitted photons is independent of the intensity of the incident radiation but increases with its frequency. And *secondly* that an increase in the intensity of radiation means an increase in the number of photons striking the plate per second and therefore an increase in the number of electrons emitted which have the same velocity and hence the same energy.

These predictions offer a satisfactory explanation of the

two experimental laws stated above and were confirmed in 1916 by Millikan's measurements and determination of Planck's constant 'A' from the photo-electric effect.

Millikan's verification of Einstein's equation. Alkali metal surfaces placed in a high vacuum tube were subjected to the action of light of known frequency, and maximum velocity v of the emitted electrons was measured by finding the potential V just needed to prevent any electrons from reaching a metal cylinder near the plate exposed to the incident radiation. Then

$$E = \frac{1}{2} mv^2 = Ve \qquad \text{...(6)}$$

and so equation (5) may be written as

$$Ve = h\nu - \omega.$$

Millikan found that graph between V and v was a straight line. The relation also indicates that the energy of the photo electrons becomes zero for a *sharply defined* frequency limit

$$\nu_0 = \omega/h,$$

thus confirming Einstein's theory.

The photoelectric cell. The photoelectric effect has numerous applications in the various fields concerning our everyday life, and the basic unit in all these cases is one or the other form of the photoelectric cell. It consists of a sealed glass bulb completely evacuated or containing one of the inert gases usually argon at low pressure. In this bulb there are two electrodes—one which forms the cathode K is an alkali metal, potassium or caesium spread on the inside wall of the glass bulb. The ether electrode which forms the anode A is a simple metallic ring inserted in the middle of the glass bulb. The two electrodes are connected to a battery B through a resistor R by means of wires, sealed into the glass bulb. When light is allowed to fall on the cathode through a window W, photoemission takes place. Under the influence of the applied potential difference these electrons travel towards the anode and a current passes through the cell. The magnitude of this photoelectric current is proportional to the intensity of light falling on the cathode so that when light of varying intensity falls on it, a fluctuating current is sent through the resistance R so that a varying potential difference is set up across its ends.

When the bulb is completely evacuated the response of the cell is *linear,* the current being proportional to the intensity of light. But if the cell contains some inert gas at low pressure and potential difference applied is not much less than that necessary to produce an ordinary gas discharge through the cell, the photo electrons are accelerated to produce ionization by collision, and the current can thus be multiplied many-fold. Such a response resulting in increase of sensitivity is said to be *non-linear* and makes this type of cell commercially important. These small photo-currents are further amplified with the aid of appropriate circuits, utilizing thermionic valves, so that very weak light signals can produce measurable effects suck cells are now being used in sound reproduction from films, picture transmission, television, traffic controls, etc. Two important modifications of the photoelectric cell are worth mentioning :

(*i*) *Solar batteries.* The photoelectric cell can be employed for producing a strong current by using a plate of pure silicon. When light falls on this, some of the electrons in its atoms are set free and tend to escape. If the surface of the plate is treated with boron, it attracts more electrons and so the free ones beneath it move up. Their movement constitutes an electric current and if the upper and lower surfaces are joined by a wire, a current will flow through it. This current is small and a strong current can be obtained by using a battery of several hundred plates. Such "solar batteries" have been used to drive cars by sunshine and are employed to supply electricity to work the radio instruments in artificial satellites.

(*ii*) *Photo multiplier tubes.* They are the special phototubes developed within the past few years in which the amplification occurs inside the photo tube. They are based on the principle of *secondary emission.*

When light strikes the surface of a photosensitive plate A, it causes the ejection of photo electrons from it ; these electrons are then attracted to a metal surface called a *dynode.* High velocity electrons hitting a metal surface can cause the ejection of one or more secondary electrons from the surface.

Suppose that a photoelectron striking dynode B produces n electrons by secondary emission. These electrons in turn are directed towards a second dynode C by making its potential higher than that of dynode B. If n electrons are again ejected by second emission for each incident electron, then for each electron emitted by the photosensitive plated there are produced n^2 electrons. This process multiplies with each successive plate until finally if there are several dynodes, each at a potential higher than the preceding one, an avalanche of electrons strikes the collector plate F resulting in a strong current in the outer circuit.

To illustrate it by an example suppose there are m dynodes so that the total number of electrons reaching the collector plate is n^{m}. If $n = 10$ and the number of dynodes is 6 then $n^{m} = 10^6$. Thus there is an enormous gain over the signal strength obtained from a simple photoelectric cell.

Photo multiplier tubes used in conjunction with suitable fluorescent material, called phosphors are now being extensively used for the detection and counting of sub-atomic particles. Such devices are known as scintillation counters.

Example. *Calculate the work function in electron—volts in case of a photo electrically active substance, when $v_0 = 4.4 \times 10^{14}\ s^{-1}$, $h = 6.625 \times 10^{-34}$ J-sec and 1 ev = 1.6×10^{-19} Joules.*

From equation (4)

$$w = h\nu_0$$

$$= 6.625 \times 10^{-34}\ \text{J-s} \times 4.4 \times 10^{14}\ \text{s}^{-1}$$

$$= 6.625 \times 4.4 \times 10^{-20}\ \text{Joule}$$

$$= \frac{6.625 \times 4.4 \times 10^{-20}}{1.6 \times 10^{-19}}\ \text{ev} \ldots \left[\because 1\ \text{ev} = 1.6 \times 10^{-19}\ \text{joule}\right]$$

$$= \mathbf{1.82\ ev.}$$

Example. *The photoelectric threshold for a certain metal is 3600 Å. Determine the maximum energy of the ejected electrons by a radiation of 2000 Å.*

$$K_{max} = \frac{1}{2} mv_{max}^2 = h(\nu - \nu_0) = hc\left(\frac{1}{\lambda} - \frac{1}{\lambda_0}\right)$$

$$K_{max} = 6.625 \times 10^{-34}\text{J-s} \times 3 \times 10^8\ \text{ms}^{-1} \times \left(\frac{10^{10}}{2000\ \text{m}} - \frac{10^{10}}{3600\ \text{m}}\right)$$

$$= 19.875 \times 10^{-16}\ \text{J-m} \times \frac{1600\ \text{m}}{2000 \times 3600\ \text{m}^2}$$

$$= \mathbf{4.416 \times 10^{-19}\ Joule}$$

Example. *What is meant by electrons volt ? Calculate in ev the energy of (a) an electron moving with a velocity of $10^7\ ms^{-1}$; (b) an α-particle with a velocity of $10^7\ ms^{-1}$, (c) a photon of sodium light of wavelength 5890×10^{-10} m.*

The electron-volt (ev) is the quantity of energy gained by an electron in falling through a potential difference of 1 volt.

Since electronic charge,

$$e = 1.6 \times 10^{-19} \text{ coulomb}$$

$$\therefore \quad 1 \text{ ev} = 1.6 \times 10^{-19} \text{ coulomb} \times 1 \text{ volt}$$

$$= 1.6 \times 10^{-19} \text{ joule.}$$

(*a*) The kinetic energy of an electron

$$E_{kin} = \frac{1}{2} mv^2$$

Putting the values in SI units, of

$$m = 9.11 \times 10^{-31} \text{ kg}$$

and $v = 10^7 \text{ ms}^{-1}$, we have

$$E_{kin} = \frac{1}{2} \times 9.11 \times 10^{-31} \text{ kg} \times (10^7 \text{ ms}^{-1})^2$$

$$= \frac{1}{2} \times 9.11 \times 10^{-31} \times 10^{14} \text{ Joule}$$

$$= 4.55 \times 11^{-17} \text{ joule}$$

$$= \frac{4.55 \times 10^{-17} \text{ J}}{1.6 \times 10^{-19} \text{ J/ev}} = \mathbf{284 \text{ ev.}}$$

(*b*) Mass of an α-particle,

$$M_\alpha = 6.64 \times 10^{-27} \text{ kg}$$

$$\text{K.E.} = \frac{1}{2} M_\alpha v^2$$

$$= \frac{1}{2} \times 6.64 \times 10^{-27} \text{ kg} \times (10^7 \text{ m s}^{-1})^2$$

$$= 3.32 \times 10^{-12} \text{ joule}$$

$$= \frac{3.32 \times 10^{-12} \text{ J}}{1.6 \times 10^{12} \text{ J/ev}} = \mathbf{2.08 \times 10^6 \text{ ev.}}$$

(*c*) The energy of a quantum of light

$$E = h\nu$$

where h is Planck's constant and ν the frequency of light.

Now, $h = 6.625 \times 10^{-34}$ J-s, $\nu = c/\lambda$,

where c, the velocity of light = 3×10^8 m s^{-1}

and wavelength of sodium light in SI units,

$$\lambda = 5890 \times 10^{-10} \text{ m.}$$

$$\therefore \quad \nu = \frac{c}{\lambda} = \frac{3 \times 10^8 \text{ m s}^{-1}}{5890 \times 10^{-10} \text{ m}}$$

$$= \frac{3}{5890} \times 10^{12} \text{ s}^{-1}$$

$$\therefore \quad E = 6.625 \times 10^{-34} \text{ J-s} \times \frac{3}{5890} \times 10^{18} \text{ s}^{-1}$$

$$= \frac{6.625 \times 3 \times 10^{-16} \text{J}}{5890}$$

$$= \frac{6.625 \times 3 \times 10^{-16} \text{J}}{5890 \times 1.6 \times 10^{-10} \text{ J/ev}}$$

$$= \frac{19.875 \times 10^2}{5890 \times 1.6} \textbf{ ev = 2.1 ev.}$$

Example. *Calculate the work function in electron volts in case of a photo-electrically active substance when* $v_0 = 4.5 \times 10^{14}$ *Hz,* $h = 6.625 \times 10^{-34}$ *J-s and 1 electron-volt (ev) =* 1.6×70^{-19} *joule.*

The work function

$$w = h\nu_0$$

$$= 6.625 \times 10^{-34} \text{ J-s} \times 4.5 \times 10^{14} \text{ s}^{-1}$$

$$= 6.625 \times 4.5 \times 10^{-20} \text{ J}$$

$$= \frac{6.625 \times 4.5 \times 10^{-20} \text{ J}}{1.6 \times 10^{-19} \text{ J/ev}}$$

... [$\because$ 1 ev = 1.6×10^{-19} joule]

$$= \textbf{1.86 ev.}$$

Example. *Radiation falls on a target within a solenoid of 20 turns per cm carrying a current of 2.5 amp. The emitted electrons then move in circles the largest of which is of radius 1 cm. Find the frequency of the incident radiation assuming the following data :*

$$e/m = 1.759 \times 10^{11} \text{ coulombs kg}^{-1} ;$$

$$h = 6.625 \times 10^{-34} \text{ J-s} ;$$

$$m = 9.11 \times 10^{-31} \text{ kg} ;$$

and $\nu_0 = 0$, *where* ν_0 *is the threshold frequency.*

Since work function w $(= h\nu_0)$ is zero, the entire energy of the incident photon is imparted to the ejected election, which moves out with velocity v, given by

$$\frac{1}{2} mv^2 = h(\nu - \nu_0) = h\nu. \quad \text{... (i)}$$

By the application of the magnetic field the electron describes a circle of radius r. If B be the flux density of the field and m the mass of the electron then,

$$Bev = \frac{mv^2}{r}$$

whence $$v = \frac{Ber}{m}.$$

Putting this value of v in (*i*), we get

$$\frac{1}{2} m \left(\frac{Ber}{m}\right)^2 = h\nu$$

Simplifying, we get

$$\nu = \frac{1}{2} \frac{m}{h} (Br)^2 \left(\frac{e}{m}\right)^2 \quad \text{... (ii)}$$

The field inside the solenoid,

$$B = \mu_0 In,$$

where $\mu_0 = 4\pi \times 10^{-7}$ Hm^{-1}, $I = 2.5$ A, and number of turns,

$$n = 2000 \text{ m}^{-1}$$

$$\therefore \quad B = 4\pi \times 10^{-7} \text{ H m}^{-1} \times 2.5 \text{ A} \times 2000 \text{ m}^{-1}$$

$$= 2\pi \times 10^{-3} \text{ Wb m}^{-2} \ldots [\because \text{HA} = \text{V-s} = \text{Wb}]$$

$$r = 1 \text{ cm} = 10^{-2} \text{ m.}$$

Substituting the data in relation (*ii*), we get

$$\nu = \frac{1}{2} \cdot \frac{9.11 \times 10^{-31} \text{ kg}}{6.625 \times 10^{-34} \text{ J-s}} \cdot$$

$(2\pi \times 10^{-3} \text{ Wb m}^{-2} \times 10^{-2} \text{ m})^2 \times (1.759 \times 10^{11} \text{ C kg}^{-1})^2$,

which on simplification gives

$$\nu = \frac{9.11 \times 4\pi^2 \times (1.759)^2 \times 10^{15}}{13.25} \text{ Hz}$$

$$= \mathbf{8.39 \times 10^{16} \text{ Hz.}}$$

Example. *The threshold frequency for photoelectric emission in a certain metal is 1.1 × 10¹⁵ Hz. Find the maximum energy of the photo electrons when light of frequency 1.5 × 10¹⁵ Hz is incident on the metal surface.*

Since $E_{max} = h(\nu - \nu_0)$

where $h = 6.625 \times 10^{-34}$ J-s

and $v_0 = 1.1 \times 10^{15}$ Hz,

$v = 1.5 \times 10^{15}$ Hz

Substituting, we have

$$E_{max} = 6.625 \times 10^{-34} \text{ J-s}$$

$$(1.5 \times 10^{15} - 1.1 \times 10^{15})\text{s}^{-1}$$

$$= 6.625 \times 10^{-34} \times 0.4 \times 10^{15} \text{ J}$$

$$= 2.65 \times 10^{-12} \text{ J}$$

Expressing it in electron-volts,

$$E_{max} = \frac{2.65 \times 10^{-19} J}{1.6 \times 10^{-19} J/ev}$$

$$= \mathbf{1.656\ ev.}$$

Example. *The work function of sodium is 2.3 ev. What is the maximum wavelength of light that will cause photo electrons to be emitted from the metal ? What will the maximum energies of the photo electrons be if 2000 Å light falls on the metal surface ?*

(*i*) Here the work function

$$w = 2.3 \text{ ev}$$

$$= 2.3 \times 1.6 \times 10^{-19} \text{ J}$$

since $w = h\nu_0$,

where v_0 is the threshold frequency,

and $h = 6.625 \times 10^{-34}$ J-s

$$\therefore \quad \nu_0 = \frac{w}{h} = \frac{2.3 \times 1.6 \times 10^{-19} \text{J}}{6.625 \times 10^{-34} \text{ J-s}}$$

$$= \frac{2.3 \times 1.6 \times 10^{15} \text{s}^{-1}}{6.625}.$$

and

$$\lambda_{max} = \lambda_0 = \frac{c}{\nu_0}$$

$$= \frac{3 \times 10^8 \text{ ms}^{-1} \times 6.625}{2.3 \times 1.6 \times 10^{15} \text{ s}^{-1}}$$

$$= \frac{19.875 \times 10^{-7}}{3.68} \text{ m}$$

$$= 5.4 \times 10^{-7} \text{ m} = \mathbf{5400\ Å.}$$

...[1 m = 10^{10}Å]

(*ii*) The maximum energy of the photo-electrons,

$$E_{max} = h(\nu - \nu_0)$$

$$= hc\left(\frac{1}{\lambda} - \frac{1}{\lambda_0}\right) = hc\frac{\lambda_0 - \lambda}{\lambda\lambda_0}$$

Here $\lambda = 2000 \text{ Å} = 2000 \times 10^{-10}$ m

$\lambda = 5400 \text{ Å} = 5400 \times 10^{-10}$ m

$$\therefore \quad E_{max} = 6.625 \times 10^{-34} \text{ J-s} \times 3 \times 10^{8} \text{ m s}^{-1}$$

$$\times \frac{(5400 - 2000)\,10^{-10}\,m}{5400 \times 2000 \times 10^{-20} m^2}$$

$$= 19.875 \times 10^{-26} \times \frac{3400 \times 10^{10}}{5400 \times 2000} \text{ J}$$

$$= \frac{19.875 \times 10^{-19}}{1.6 \times 10^{-19} \text{ J/ev}} \times \frac{43 \text{ J}}{54 \times 2}$$

$$= \frac{19.875 \times 17}{1.6 \times 54} \textbf{ ev} = \textbf{3.9 ev.}$$

Example. *If sodium surface in vacuum is illuminated with a monochromatic beam of ultraviolet light with a wavelength of 2×10^{-7} m. What is the maximum velocity of emission of the function of sodium is 2.46 electron-volts ?*

From the equation, $\frac{1}{2} mv^2_{max} = h\nu - w$ ... *cf.*Eq. 5

$$h\nu = \frac{hc}{\lambda} = \frac{6.625 \times 10^{-34} \text{ J-s} \times 3 \times 10^{8} \text{ m s}^{-1}}{2 \times 10^{-7} \text{ m}}$$

$$= \frac{6.625 \times 3 \times 10^{-19}}{2} \text{ J}$$

and $w = 2.46 \text{ ev} = 2.46 \times 1.6 \times 10^{-19}$ J

$$\therefore \quad \frac{1}{2} mv_{\max}^2 = \frac{6.625 \times 3 \times 10^{-19}}{2} J - 2.46 \times 1.6 \times 10^{-19} J$$

$$= (9.937 \times 10^{-19} - 3.936 \times 10^{-18}) \text{ J}$$

$$= 6.00 \times 10^{-19} \text{ J}$$

$$\therefore \quad v_{max} = \sqrt{\frac{2 \times 6.00 \times 10^{-19} J}{9.11 \times 10^{-31} kg}}$$

$$= 10^6 \sqrt{\frac{12.0}{9.11}} \text{ ms}^{-1} = 1.15 \times 10^6 \text{ ms}^{-1}.$$

Example. *The threshold wavelength for photoelectric emission in tungsten is 2300 Å. What wavelength of light must be used so that electrons with a maximum energy of 1.5 ev are ejected.*

Since, $E_{max} = \frac{1}{2} mv_{\max}^2 = h(\nu - \nu_0)$

$$\nu = \frac{E_{max}}{h} + \nu_0 \qquad \dots (i)$$

so that $E_{max} = 1.5 \text{ ev} = 1.5 \times 1.6 \times 10^{-19} \text{ J}$

$c = 3 \times 10^8 \text{ ms}^{-1}$

$$\nu_0 = \frac{c}{\lambda_0} = \frac{3 \times 10^6 \text{ m s}^{-1}}{2300 \times 10^{-10} \text{ m}}$$

$$= \frac{30}{23} \times 10^{15} \text{ Hz}$$

$h = 6.625 \times 10^{-34}$ J-s

$\nu = ?$

Substituting in (*i*), we get

$$v = \frac{1.5 \times 1.6 \times 10^{-19} J}{6.625 \times 10^{-34} J\text{-}s} + \frac{30}{23} \times 10^{15} \text{ Hz}$$

$$= \frac{2.40}{6.625} \times 10^{15} \text{ Hz} + \frac{30}{23} \times 10^{15} \text{Hz}$$

$$= \left[\frac{2400}{6625} + \frac{30}{23}\right] \times 10^{15} \text{ Hz}$$

$$= \left[\frac{96}{265} + \frac{30}{23}\right] \times 10^{15} \text{ Hz}$$

$$= \left[\frac{96 \times 23 + 30 \times 265}{265 \times 23}\right] \times 10^{15} \text{ Hz}$$

$$= 1.667 \times 10^{15} \text{ Hz}$$

Now $$\lambda = \frac{c}{v} = \frac{3 \times 10^{8} \text{ ms}^{-1}}{1.667 \times 10^{15} \text{ s}^{-1}}$$

$$= 1.8 \times 10^{-7} \text{ m} = \mathbf{1800 \text{ Å}}.$$

Origin of line spectra (Bohr's theory of the hydrogen spectrum). A striking evidence in favour of Planck's quantum hypothesis was furnished by Niels Bohr's work in 1913 regarding satisfactory explanation of the origin of line spectra of hydrogen and other elements. Even today Bohr's theory of line spectra forms the backbone of modern physics.

According to Rutherford's model of the atom there should be an infinite number of possible orbits of a circulating electron, as the only condition to be satisfied on the classical theory is that the centripetal force $\frac{mv^2}{r}$ ($= mr\omega^2$) is to be provided by the attractive force between the nucleus and the electron, that is

$$mr\omega^2 = \frac{Ze \times e}{r^2} \quad \text{... } (i)$$

where Z is the atomic number of the nucleus, m the mass of the revolving electron, r its radius and ω its angular velocity.

Note. In SI units the attractive force between the nucleus and the electron is $\frac{1}{4\pi \epsilon_0} \times \frac{Ze^2}{r^2}$, so that equation (*i*) is written as

$$mr\omega^2 = \frac{1}{4\pi \epsilon_0} \times \frac{Ze^2}{r^2} \qquad \text{... }(ii)$$

Again an electron moving in a circle possesses acceleration and on the basis of classical electromagnetic theory it should radiate energy continuously following a spiral path towards the nucleus and may ultimately result in the collapse of the atom. Thus the atom would be both indefinite and unstable and could not thus exist. But this is not so. Moreover, the continuous drop in energy of the revolving electron should also result in a continuous spectrum with all possible wavelengths present, in direct contradiction of the experimental observations on the hydrogen spectrum which shows well-defined spectral lines. This was the situation in 1911 when Niels Bohr came to Manchester. It was then that, impressed by his sharp intellect, Rutherford suggested to him to turn to the theoretical aspect of the problem.

Application of quantum theory. Soon after in 1913 Bohr applied the quantum theory to Rutherford model of the atom making a bold departure from classical principles. He assumed that :

(*i*) a revolving electron did not radiate energy continuously;

(*ii*) it emitted or absorbed either a quantum of energy $h\nu$ or no energy at all thus ensuring the stability of the atom;

(*iii*) to eliminate the possibility of indefiniteness of the atom he postulated that electron had certain stationary or non-radiating orbits for which the angular momentum of the electron is an integral multiple of $h/2\pi$, where h is Planck's constant ; and (*iv*) energy is radiated only when an electron takes a spontaneous jump from an outer to an inner orbit.

Now the angular momentum of a particle of mass *m* moving with a speed *v* in a circle of radius *r* is *mvr*. The postulate (*iii*) may then be stated as

$$mvr = n\frac{h}{2\pi} = n\hbar$$

$$\left[\frac{h}{2\pi} = \hbar = 1.0546 \times 10^{-34} \text{ J - s}\right]$$

where *n* is an integer and $\hbar$ is known as unit angular momentum.

Since $v = rw$, we have

$$mr^2w = n\frac{h}{2\pi} \qquad \text{... } (iii)$$

Solving (*ii*) and (*iii*) we get the radii of the permissible orbits

$$r_n = \frac{n^2h^2 \times 4\pi\varepsilon_0}{4\pi^2 mZe^2} = \frac{\varepsilon_0 n^2 h^2}{\pi m Z e^2} \qquad \text{... (7)}$$

This expression shows that the radii of the stable orbits in which the electron can circulate without losing energy are proportional to 1^2, 2^2, 3^2, etc. Thus the radius of the inner-most orbit ($n = 1$) of hydrogen is given by

$$r_1 = \frac{\varepsilon_0 h^2}{\pi m e^2}$$

Substituting the known values of

$$e_0 = 8.85 \times 10^{-12} \frac{\text{Coulomb}^2}{\text{Newton - m}^2}$$

$$h = 6.625 \times 10^{-34} \text{ J-s} ;$$

$$m = 9.11 \times 10^{31} \text{ kg} ;$$

$$e = 1.601 \times 10^{-19}\ \text{C};$$

$$Z = 1, \text{ we find that};$$

$$r_1 = \frac{8.85 \times 10^{-12} \times \left(6.625 \times 10^{-34}\right)^2}{\pi \times 9.11 \times 10^{-31} \times \left(1.601 \times 10^{-19}\right)^2}$$

$$= 0.528 \times 10^{-11}\ \text{m}.$$

The result agrees remarkably well with the value obtained from kinetic theory of gases. Thus, in the case of hydrogen atom in the normal state, the electron occupies the smallest of the permissible stationary orbits.

The other radii are given in terms of r_1 by the relation

$$r_n = n^2 r_1, \ n = 1, 2, 3, \ldots\ldots \qquad \ldots (7a)$$

So that the spacing between adjacent orbits increases progressively. The success of Bohr's theory lay in the explanation of hydrogen spectrum as the spectral formula theoretically derived was found to be in good accord with the observed spectral values.

Derivation of spectral formula. In the case of the hydrogen atom there is a single electron revolving round the proton. The total energy of the electron in the orbit is the sum of its kinetic and potential energies.

The kinetic energy $= \frac{1}{2} mv^2 = \frac{1}{2} mr^2 \omega^2$. Considering the electrical potential to be zero at infinite distance from the change $+e$ of the nucleus, the potential at a distance r is $+e/\pi \varepsilon_0 r$ and the potential energy for the electronic charge $-e$ is

$$-e \times \frac{e}{4\pi\varepsilon_0 r} = -\frac{e^2}{4\pi\varepsilon_0 r}$$

$\therefore$ total energy of the electron,

$$E = \frac{1}{2} mr^2 \omega^2 - \frac{e^2}{4\pi\varepsilon_0 r} \qquad \ldots (iv)$$

But from equation (*ii*),

$$mr\omega^2 = \frac{e^2}{4\pi\varepsilon_0 r}$$

or
$$mr^2\omega^2 = \frac{e^2}{8\pi\varepsilon_0 r} \qquad \ldots [\because Z = 1]$$

$\therefore$
$$E = \left(\frac{e^2}{8\pi\varepsilon_0 r} - \frac{e^2}{4\pi\varepsilon_0 r}\right) = -\frac{e^2}{8\pi\varepsilon_0 r}$$

Substituting the value of r from equation (7), we get

$$E_n = -\frac{e^2}{8\pi\varepsilon_0} \times \frac{\pi me^2}{\varepsilon_0 n^2 h^2} = \frac{me^4}{8\varepsilon_0^2 n^2 h^2}. \qquad \ldots (v)$$

The total energy has a negative sign because the reference level of potential energy is taken with the electron at an infinite distance from the nucleus.

As n increases E also increases since E is negative. This means that the energy in an outer orbit is greater than that in the inner orbit. According to the fourth assumption made by Bohr the energy is radiated only when an electron takes a spontaneous jump from an outer to the inner orbit. The radiated energy was assumed to be related to the frequency of light by the quantum formula $E = h\nu$.

Thus if E_2 is the energy in the outer orbit and E_1 in the inner

$$E_2 - E_1 = h\nu. \qquad \ldots (vi)$$

The question arises, wherefrom this energy comes initially? The electron normally revolving in the smallest orbit must first be raised to a higher orbit. This can be accomplished in an electric discharge tube by the impact of the fast moving electrons through the gas. In any event, once the electron is elevated to one of the larger permissible orbits, it is in a position to fall back spontaneously into a smaller orbit. It may not go to the smallest orbit in a single jump but may do so in several steps. Each transition produces a spectral line given by

$$E = h\nu,$$

where E is the difference in the energy of the initial and final orbits involved in the particular jump. So that we can write energy of the emitted photon = energy of atom before emission – energy of atom after emission.

If an electron jumps from an orbit n_2 to an orbit n_1 we have from equation (*vi*) above

$$h\nu = E_2 - E_1 = \frac{me^4}{8\varepsilon_0^2 h^2} - \left(\frac{1}{n_1^2} - \frac{1}{n_2^2}\right)$$

$$\therefore \qquad \nu = \frac{me^4}{8\varepsilon_0^2 h^3} - \left(\frac{1}{n_1^2} - \frac{1}{n_2^2}\right) \qquad \text{... (8)}$$

Comparison of this equation with Balmer's series expressed in equation (1), shows that the two are identical if

$$R = \frac{me^4}{8\varepsilon_0^2 h^3} \qquad \text{... (9)}$$

and $\qquad n_1 = 2$ and $n_2 = 3, 4, 5, \ldots \ldots$

If we substitute known values of $m = 9.1 \times 10^{-91}$ kg, $e = 1.6 \times 10^{-19}$ C and $h = 6.625 \times 10^{-34}$ J-s in equation (9) we obtain for R – a value = 3.291×10^{15} s^{-1} in exact agreement with the observed value within the limits of experimental error.

Other series indicated in equations (3), (4), (5) and (6) are also obtained from equation (8) above.

Putting $n_1 = 1$, $n_2 = 2, 3, 4\ldots$ we have Lyman series.

If we make $n_1 = 3$ and 4 and $n_2 = 4, 5, 6$, etc. and 5, 6, 7 etc., respectively we get Ritz-Paschen and Brackett series.

For Pfund series $n_1 = 5$ and $n_2 = 6, 7, 8$ etc.

The Energy in an orbit. Let us now calculate the energy of nth quantum orbit of the hydrogen atom in the SI* units.

The total energy in any orbit, say *n*th, is the sum of its kinetic and potential energies.

The kinetic energy $(E_{kin}) = \frac{1}{2} m r_n^2 \omega^2$ ... (9)

The potential energy of the electron when it is at a distance r_n from the nucleus is *equal* and *opposite* to the work done in removing the electron to infinity against the electrostatic force of attraction between the nucleus and the electron, the infinity being the seat of zero energy.

Now the pull on the electron,

$$F_e = \frac{1}{4\pi\varepsilon_0} \times \frac{Ze^2}{r^2}$$

$$= \frac{e^2}{4\pi\varepsilon_0 r^2} \quad \text{... [Z = 1]}$$

... $[e_0 = 8.854 \times 10^{-12} \text{ Fm}^{-1}]$

Hence potential energy

$$E_{pot} = -\int_{r_n}^{\infty} F_e dr$$

$$= -\int_{r_n}^{\infty} \frac{e^2}{4\pi\varepsilon_0 r^2} dr$$

$$= -\left|\frac{e^2}{4\pi\varepsilon_0 r}\right|_{\infty}^{r_n} = -\frac{e^2}{4\pi\varepsilon_0 r_n} \quad \text{... (10)}$$

∴ total energy in the *n*th orbit,

$$E_n = E_{kin} + E_{pot}$$

$$= \frac{1}{2} m r_n^2 \omega^2 - \frac{e^2}{4\pi\varepsilon_0 r_n}$$

From Equation (*ii*),

$$mr_n\omega^2 = \frac{1}{4\pi\varepsilon_0 r_n} \times \frac{Ze^2}{r_n^2}$$

or $$mr_n^2\omega^2 = \frac{e^2}{4\pi\varepsilon_0 r_n} \qquad \text{... [Z = 1]}$$

$$\therefore \qquad E_n = \frac{e^2}{8\pi\varepsilon_0 r_n} - \frac{e^2}{4\pi\varepsilon_0 r_n}$$

$$= -\frac{e^2}{8\pi\varepsilon_0 r_n} \qquad \text{... (10}a\text{)}$$

Substituting the value of $r_n \left(= \frac{\varepsilon_0 n^2 h^2}{\pi m e^2}\right)$ obtained from the solution of equations (*i*) and (*ii*), that is, $mr^2\omega = \frac{nh}{2\pi}$ and $mr\omega^2 = \frac{1}{4\pi\varepsilon_0} \times \frac{e^2}{r^2}$, we get

$$E_n = \frac{e^2}{8\pi\varepsilon_0} \times \frac{\pi m e^2}{\varepsilon_0 n^2 h^2}$$

$$= -\frac{me^4}{8\varepsilon_0^2 h^2}\left(\frac{1}{n^2}\right), \; n = 1, 2, 3, \text{........} \qquad \text{(11)}$$

For an element with nuclear charge *Ze*, Equation (11) becomes

$$E_n = -\frac{Z^2 me^4}{8\varepsilon_0^2 h^2} \cdot \left(\frac{1}{n^2}\right) \qquad \text{... (12)}$$

for in this case e^2 is replaced by Ze^2.

Equation (12) may be used to calculate energy of the electron in the different orbits of hydrogen atom.

Thus the energy in the ground orbit

$$E_1 = -\frac{me^4}{8\varepsilon_0^2 h^2} \cdot \left(\frac{1}{1^2}\right) \qquad \text{... (12a)}$$

$$= -\frac{9.11 \times 10^{-31}\, kg \times \left(1.6 \times 10^{-19} C\right)^4}{8 \times \left(8.854 \times 10^{-12}\, Fm^{-1}\right)^2 \times \left(6.625 \times 10^{-24}\, J\text{-}s\right)^2}$$

Simplification gives

$$E_1 = \frac{9.11 \times (1.6)^4 \times 10^{-15}}{8 \times (8.854)^2 \times (6.625)} \text{ J} \quad [\because C^4 = F^2J^2]$$

$$= -21.76 \times 10^{-19} \text{ J} \qquad \text{... (13)}$$

In the next higher orbit

$$E_2 = \frac{E_1}{2^2}$$

$$= -\frac{21.76 \times 10^{-19} \text{ J}}{4}$$

$$= -5.44 \times 10^{-19} \text{ J.}$$

Similarly, $$E_3 = \frac{E_1}{3^2}$$

$$= -\frac{21.76 \times 10^{-19} \text{ J}}{9}$$

$$= -2.42 \times 10^{-19} \text{ J}$$

and so on.

If an excited atom has its electron fall from an orbit n_2 to n_1, then

Initial energy, $$E_2 = -\frac{me^4}{8\varepsilon_0^2 h^2} \cdot \frac{1}{n_2^2}$$

Final energy, $$E_1 = -\frac{me^4}{8\varepsilon_0^2 h^2} \cdot \frac{1}{n_1^2}$$

Hence the energy difference between these states is

$$E_2 - E_1 = -\frac{me^4}{8\varepsilon_0^2 h^2}\left[\left(-\frac{1}{n_2^2}\right)\right]-\left[\left(-\frac{1}{n_1^2}\right)\right]$$

$$= -\frac{me^4}{8\varepsilon_0^2 h^2}\left(\frac{1}{n_1^2}-\frac{1}{n_2^2}\right) \quad \text{... (14)}$$

The frequency ν of the photon released in the transition is given by

$$E_2 - E_1 = h\nu$$

or

$$\nu = \frac{E_2 - E_1}{h}$$

$$= -\frac{me^4}{8\varepsilon_0^2 h^3}\left(\frac{1}{n_1^2}-\frac{1}{n_2^2}\right) \quad \text{... (15)}$$

In terms of photon wavelength λ, we have $\frac{1}{\lambda} = \frac{v}{c}, \frac{1}{\lambda}$ called the wave number and is denoted by $\bar{\nu}$.

$$\therefore \quad \bar{\nu} = \frac{me^4}{8\varepsilon_0^2 ch^3}\left(\frac{1}{n_1^2}-\frac{1}{n_2^2}\right) \quad \text{... (16)}$$

Equation (16) states that the radiation emitted by hydrogen atom should contain certain wavelengths only.

Comparison of this equation with Balmer's series expressed in Equation (7), shows that the two are identical, if the Rydberg constant

$$R = \frac{me^4}{8\varepsilon_0^2 ch^3} \quad \text{... (17)}$$

and $\quad n_1 = 2$

and $\quad n_2 = 3, 4, 5, \ldots$

If we substitute known values

$$m = 9.11 \times 10^{-31} \text{ kg};$$

$$e = 1.6 \times 10^{-19} \text{ C};$$

$$\varepsilon_0 = 8.854 \times 10^{-12} \text{ Fm}^{-1};$$

$$c = 3 \times 10^{8} \text{ ms}^{-1};$$

and

$$h = 6.625 \times 10^{-34} \text{ J-s}.$$

in equation (17), we have

$$R = \frac{me^4}{8\varepsilon_0^2 ch^3}$$

$$= \frac{9.11 \times 10^{-31} \text{ kg} \times \left(1.6 \times 10^{-19} \text{C}\right)^4}{8 \times \left(8.854 \times 10^{-12} \text{ Fm}^{-1}\right)^2 \times 3 \times 10^8 \text{ ms}^{-1} \times \left(6.625 \times 10^{-34} \text{ J-s}\right)^2}$$

Simplification yields

$$R = \frac{9.11 \times (1.6)^4 \times 10^{11}}{8 \times (8.854)^2 \times 3 \times (6.625)^2} \text{ m}^{-2}. \qquad [\because C_4 = F^2J^2]$$

$$= \mathbf{1.091 \times 10^7 \text{ m}^{-1}} \qquad \dots (18)$$

in fair agreement with the observed value within the limits of experimental error [see Equation (2)].

Example. *Calculate on Bohr's theory, (a) the velocity (b) the energy of an electron in the innermost orbit of the hydrogen atom using SI unit.*

(*a*) The velocity of the electron in the ground orbit may be calculated from the equation

$$\frac{mv^2}{r_1} = \frac{1}{4\pi\varepsilon_0} \cdot \frac{e^2}{r_1^2}$$

$$v^2 = \frac{1}{4\pi\varepsilon_0 m} \cdot \frac{e^2}{r_1} \qquad \dots (i)$$

Now $r_1 = \frac{\varepsilon_0 h^2}{\pi m e^2}$, ... [since $n = 1$] ... (*ii*)

∴ substituting the value of r_1 in (*i*)

$$v^2 = \frac{1}{4\pi\varepsilon_0 m} \cdot \frac{e^2 \pi m e^2}{\varepsilon_0 h^2}$$

$$= \frac{e^4}{4\varepsilon_0^2 h^2},$$

$$\text{whence } v = \sqrt{\frac{e^4}{4\varepsilon_0^2 h^2}}$$

$$= \frac{e^2}{2\varepsilon_0 h}$$

$$= \frac{\left(1.6 \times 10^{-10}\ C\right)^2}{2 \times 8.854 \times 10^{-12}\ Fm^{-1} \times 6.625 \times 10^{-34}\ J.5}$$

$$= \frac{(1.6)^2 \times 10^{-33}}{2 \times 8.854 \times 6.625 \times 10^{-44}}\ \text{ms}^{-1} \quad \text{... } [\because C^2 = \text{FJ}]$$

$$= \mathbf{2.18 \times 10^6\ m\ s^{-1}}$$

(*b*) The energy, E_1 of the electron in the ground state is obtained by setting the values of the constants m, ε_0, h, e, and $n = 1$ in Equation (12 *a*).

$$E_1 = -\frac{me^4}{8\varepsilon_0^2 h^2}\left(\frac{1}{1^2}\right)$$

$$= -\frac{9.11 \times 10^{-31}\ \text{kg} \times \left(1.60 \times 10^{-19}\ \text{C}\right)^4}{8 \times \left(8.854 \times 10^{-12}\ \text{Fm}^{-1}\right)^2 \times \left(6.625 \times 10^{-34}\ \text{J-s}\right)^2}\left(\frac{1}{1^2}\right)$$

$$= -\frac{9.11 \times (1.6)^4 \times 10^{-15}}{8 \times (8.854)^2 \times (6.625)^2}\ \text{J} \qquad [\because C^4 = F^2J^2]$$

$$= \mathbf{-21.76 \times 10^{-10}\ joule}$$

Example. *The lowest two excited states of a hydrogen atom are 10.2 and 12.0 ev above the ground state. Calculate the ware lengths of radiation that could be produced by transition between these states and the ground state.*

[Take $h = 6.62 \times 10^{-34}$ *J-s and* $c = 3\times 10^{8}$ *m* s^{-1}]

(*i*) Energy emitted in going from 1st excited state to the ground state.

$$E_1 = 10.2 \text{ ev}$$

Now,

$$E_1 = hv$$

$$= \frac{hc}{\lambda_1}.$$

where λ_1 is the wavelength of the radiation produced,

$$\therefore \qquad \frac{hc}{\lambda_1} = 10.2 \text{ ev}$$

or

$$\lambda_1 = \frac{hc}{10.2 \text{ ev}}$$

$$= \frac{6.62 \times 10^{-34} \; j\text{-}s \times 3 \times 10^{8} \; m\text{-}s^{-1}}{10.2 \times 1.6 \times 10^{-19} J}$$

$$= \frac{19.86}{16.32} \times 10^{-7} \text{ m}$$

$$= \mathbf{1.22 \times 10^{-7} \text{ m}}.$$

(*ii*) Similarly, the wavelength of the radiation produced in going from 2nd excited state to ground state is given by

$$\lambda_2 = \frac{hc}{12.0 \text{ ev}}$$

$$= \frac{6.62 \times 10^{-34} \text{ J-s} \times 3 \times 10^{8} \text{ ms}^{-1}}{12.0 \times 1.6 \times 10^{-19} \text{ J}}$$

$$= \frac{19.86}{19.2} \times 10^{-7} \text{ m}$$

$$= \mathbf{1.035 \times 10^{-7} \text{ m.}}$$

(*iii*) Energy emitted during transition between the two excited states

$$E = E_2 - E_1$$

$$= (12.0 - 10.2) \text{ ev}$$

$$= \frac{hc}{\lambda_3},$$

where λ_3 is the wavelength of the emitted radiation.

$$\therefore \qquad \lambda_3 = \frac{hc}{E_2 - E_1}$$

$$= \frac{hc}{(12.0 - 10.2) \text{ ev}}$$

$$= \frac{6.62 \times 10^{-34} \text{ J-s} \times 3 \times 10^8 \text{ ms}^{-1}}{1.8 \times 1.6 \times 10^{-10} \text{ J}}$$

$$= \frac{6.62 \times 3 \times 10^{-7}}{1.8 \times 1.6} \text{ m}$$

$$= \mathbf{6.9 \times 10^{-7} \text{ m.}}$$

Interpretation of absorption spectra on Bohr's theory. If light from an arc lamp is examined by a spectroscope after it has passed through a gas at low pressure and temperature lower than that of the source, it is found that the spectrum is traversed by certain dark lines showing that certain wavelengths are absorbed. This is called the absorption spectrum of the substance Like the emission spectra, we have both the *continuous* or *general* absorption and *selective or line* absorption.

Let us examine what is the mechanism involved on Bohr's

theory. When white light from a strong source which contains all wavelengths, is passed through hydrogen gas or a sodium flame, photons of these wavelengths that correspond to transition between different energy levels are absorbed. The resulting excited hydrogen or sodium atoms re-radiate their excitation energy almost at once, but these photons come off in random direction with only a few in the same direction as the original beam of white light. The dark lines in an absorption spectrum are, therefore, never completely black, but only appear so by contrast with the bright background. We expect the absorption spectrum of any element to be identical with its emission spectrum and this agrees with our observations. For example, in the case of sodium the two black absorption lines occupy exactly the same position in the spectrum as the two yellow D_1 and D_2 emission lines.

Interpretation of hydrogen spectra. The Lyman series arises from electrons returning from the various excited states of the atom directly to the normal state. Balmer series from electrons returning to the second stationary orbit and so on for other series. The theory has been applied wish success to helium spectrum, and *X*-rays spectra. Since optical spectra in case of other atoms are complex owing to electrons in the orbits exerting mutual forces, simple theory cannot be applied. But the principle that there are energy levels or orbits still holds good and all the spectral lines are caused by the emission of a quantity of energy of the electrons in jumping from one level to the other.

Bohr's theory also explains why there are fewer lines in the absorption spectrum than in the emission spectrum. In a cool gas all the atoms will be in their normal or unexcited state. Hence only those quanta of energy or photons as they are called can be absorbed whose energies are just sufficient to raise the atoms from their normal states to one of the excited states. Therefore it is obvious that only those lines of the spectrum which are produced as a consequence of the jumping of the electrons back to the normal state will make their appearance in the absorption spectrum. Thus in the case of sodium the cool vapour will produce an absorption spectrum which consists of the principal series only.

An extension of Bohr's theory to the spectra of other complex atoms has supported our views about the origin of spectra and we have strong evidence that the ultimate source or origin from which light waves emanate must in some way be related to the atom and must be electrical in nature. This view is further supported by the *Zeeman* and *Stark effects*. In the former case a luminous source is influenced to a marked degree by a strong magnetic field and in the latter case the field is electric.

Discrete atomic energy levels—Franck and Hertz experiment. In investigating the nature of radiation we have had ample evidence concerning its corpuscular nature during its interaction with matter as in photo-electric effect and even when it preserves its existence as in the Compton effect ; it now only remains to show the existence of a corresponding *discontinuity* in the matter itself. The explanation of emission spectra in hydrogen and other atoms has postulated the existence of standing states or discrete energy levels in the atom. There is assumed to be the lowest or the ground state corresponding to the quantum number one.

The states which correspond to the quantum numbers greater than unity are called the *excited states* of the atom because their energies are greater than that of the ground state. A transition from the lowest state to an excited state can occur only when an atom receives an amount of energy equal to the excitation energy for that particular state, for example, by electron bombardment in a gas discharge tube. *The potential through which the bombarding electron must fall to raise the collided atom to one of the excited states is called resonance or excitation potential of the atom.* For each atom there are several excitation potentials. *If, however, the bombarding electron has sufficient energy to detach completely an electron from the atom or the molecule the corresponding voltage is know as* ionization potential *and the atom or the molecule is said to be* ionized. The energy of the electron is measured in electron-volts (ev). One electron-volt is the energy acquired by an electron in falling through a potential difference of one volt.

In SI system 1 ev = 1.601×10^{-19} C×1 volt = 1.601×10^{-19} J.

The connection between the term values in the spectral series and excitation energies was demonstrated experimentally by the work of Franck and Hertz (1914). The experiment established the existence *of discrete* energy states in the atom. A metallic filament F coated with barium and strontium oxides is heated with electric current to emit a stream of electrons. They are accelerated to the grid G by a known potential V passing which they reach the plate P, called the collector, at a slightly lower potential than the grid. *i.e.*, $V_P > V_G$, F, G and P are enclosed in a glass vessel containing the element under test at suitable pressure that the electrons will suffer a number of elastic collisions with the atoms of the gas between F and G and not between G and P. When the filament current is switched on and the potential V is gradually raised from 1 volt upwards, the current registered by the galvanometer M increases steadily due to the increasing velocity of the electrons until a critical value V_1 is reached when It suddem, drops. *The critical value* V_1 *is that at which the electron makes an inelastic collision with the atom giving up the whole of its energy to the atom to raise it from the ground state to a state of hither energy.* Hence the electron has now no energy left with which to overcome the small retarding potential between the grid and the plate and therefore it fails to reach the plate reducing the current to zero. This shows the existence of a state of the atom in which its energy is greater by a finite amount than that in its ground state. As V is increased further the current increases once more. Similar drops in current have been detected at potentials V_2 and so on. All these values measure the several excitation potentials.

The excited atoms on returning to the ground state give out as radiation the energy which they received from the colliding electrons during excitation. Thus when V is greater than V_1, a radiation of definite wavelength is emitted. By raising the discharge tube voltage in steps, more and more spectral lines are added. Franck and Hertz actually found that in the case of mercury vapour first excitation state occurs at 4.9 volts and a line of wavelength 2537 Å is emitted. On Bohr's prediction we can easily calculate the wavelength corresponding to an excitation potential of 4.9 volts. The voltage will accelerate the electron imparting it an energy equivalent to

$$E = 4.9 \text{ ev} = 4.9 \times 1.601 \times 10^{-19} \text{ joule}$$

$$= 7.8449 \times 10^{-19} \text{ joule.}$$

Now $$E = h\nu = hc/\lambda$$

$\therefore$ $$\lambda = \frac{hc}{E}$$

$$= \frac{6.625 \times 10^{-34} \text{ J-s} \times 3 \times 10^{8} \text{ m-s}^{-1}}{7.8449 \times 10^{-19} \text{ joule}}$$

$$= 2.533 \times 10^{-7} \text{ m}$$

$$= 2533 \text{ Å.}$$

The result is in good agreement with the experimental value of Franck and Hertz. Thus the experiment proves not only that the atom exists in a set of discrete energy states but also that the emission process is controlled by the familiar equation

$$E = h\nu,$$

where E is the amount of energy lost by the atom in changing from the excited to the ground state and ν is the frequency of the emitted radiation. Many similar results have since been obtained with a large number of gases and Bohr's relationship between the *excited states* and the *term values* of the spectral lines has been completely verified by direct experiments.

For hydrogen the values of excitation potentials have successively been found to be 10.2, 12.0, 127, 13.0 and 13.5 volts respectively. On calculating the frequencies of the lines emitted in the case of mercury vapour we find that these values agree with observed frequencies in every case. Generalising in the case of hydrogen the energy levels according to Bohr are given by

$$E = -\frac{Rh}{n}$$

where R (=3.291 × 10^{15} s^{-1}) is the Rydberg's constant and h is Planck's constant, n refers to the nth energy state.

Hence the excitation energy from the ground to the nth state is given by

$$E_n - E_1 = Rh\left(\frac{1}{1^2} - \frac{1}{n^2}\right)$$

Substituting the values in SI units

$$E_n - E_1 = 3.291 \times 10^{15}\ \text{s}^{-1} \times 6.625 \times 10^{-34}\ \text{J-s}\left(\frac{1}{1^2} - \frac{1}{n^2}\right)$$

$$= 3.291 \times 6.625 \times 10^{-19}\left(\frac{1}{1^2} - \frac{1}{n^2}\right)\ \text{joule}$$

$$= \frac{3.291 \times 6.625 \times 10^{-19}}{1.601 \times 10^{-19}}\left(\frac{1}{1^2} - \frac{1}{n^2}\right)\ \text{ev}$$

... [$\because$ 1.601 × 10^{-19} joule = 1 ev]

$$= \frac{3.291 \times 6.625}{1.601}\left(\frac{1}{1^2} - \frac{1}{n^2}\right) \text{ ev}$$

$$= 13.61\left(\frac{1}{1^2} - \frac{1}{n^2}\right) \text{ ev}$$

By putting $n = 2, 3...$ we get the successive *excitation potentials as* 10.21, 12.10... in almost perfect agreement with the observed values. By substituting ∞ for n, we get the ionization energy

$$E_\infty - E_1 = 13.61\left(\frac{1}{1^2} - \frac{1}{\infty^2}\right) \text{ ev.}$$

This corresponds to an *ionization potential* of 13.61 volts for hydrogen atom tallying remarkably well with the experimental value.

Example. *Calculate the ionization potential of hydrogen atoms.*

In the case of hydrogen, we have

$$\nu = \frac{me^4}{8\varepsilon_0^2 h^2}\left(\frac{1}{n_1^2} - \frac{1}{n_2^2}\right)$$

cf. equation (15)

For ionization of a normal atom

$$n_1 = 1$$

and $$n_2 = \infty$$

$\therefore$ energy required for ionization is equal to

$$h\nu = \frac{me^4}{8\varepsilon_0^2 h^2} \text{ joules}$$

If V be the ionization potential in volts, then

$$eV = \frac{me^4}{8\varepsilon_0^2 h^2}$$

$$V = \frac{me^3}{8\varepsilon_0^2 h^2}$$

$$= \frac{9.11\times10^{-31}\times\left(1.601\times10^{-19}\right)^3}{8\times\left(8.854\times10^{-12}\right)^2\times\left(6.625\times10^{-34}\right)^2}$$

$$= \frac{9.11\times(1.601)^3\times10^4}{8\times(8.854)^2\times(6.625)^2}$$

$$= \textbf{13.59 volt.}$$

Band spectra of molecules. When the substances are in the molecular states we get *band spectra* consisting of bright bands, the intensity of which varies from one end to the other. On close examination with a high resolving power spectrometer we find that each band consists of fine lines which are very near to each other at the sharp end of the band called the *band head* and are gradually more and more distant towards the diffuse *(tail)* end. The greater intensity of the head is usually because of the close grouping of the lines at that point, although sometimes the lines in that part are also more intense. They can be explained by the quantum principles applied to line spectra and their analysis provides us a probe into the structure of the molecules.

The molecules of each substance have characteristic energy levels and the process of radiation consists in shifts from one level to another of lower energy. Unlike the atoms there are many more energy levels (1) the irrational energy levels (E_v) due to the molecule vibrating as a whole, (2) the rotational energy levels (E_r) due to the rotation of the molecule about an axis perpendicular to the line joining the atoms in it, and (3) the electronic energy levels (E_e) due to the revolving electrons in each atom of the molecule. Then the total energy of a molecule at any time is given by

$$E = E_\nu + E_r + E_e$$

In the case of molecules, therefore, the number of possible transitions is much greater, thus giving rise to a spectrum with many more lines. Just as electronic energy levels could have only certain definite values

$$(E_e)_0, (E_e)_1, (E_e)_2 \text{ etc.}$$

so is the case with vibrational and rotational energy levels having values

$$(E_v)_0, (E_v)_1, (E_v)_2,$$

and $$(E_r)_0, (E_r)_1, (E_r)_2 \text{ etc. respectively.}$$

The molecules at any instant could have any of the values in the respective series and the process of emission of radiant energy involves the change in the value of any one or any two or all the three components of energy. In the most general case when a quantum $h\nu$ is radiated, we may write

$$h\nu = h\nu_0 + h\nu_r + h\nu_e$$

where $h\nu_0$, is the energy arising from a vibrational change in level, hv_r from a rotational change and hv_e from an electronic change.

If only E_r ($h\nu_r$) or E_v ($h\nu_v$) changes, we have a series of energy levels which are relatively close together and the radiant energy may be far out in the longer wavelengths, a region in which it is difficult to work as glass absorbs all such radiations. The major energy changes, however, take place only during the transitions in the electronic energy levels just as in the case of ordinary lino spectra, when the radiation emitted falls in the visible or the ultraviolet region.

But when the changes in the values of the rotational and vibrational energy levels are added to the above simultaneously, we should obviously expect a complex group of lines. These constitute our so called band spectra. The more complex the molecule, the more complicated the spectrum and so far we are successful in analysing only a few simpler ones

While only a brief analysis of the band spectra has been given it should be sufficient to enable the reader to understand the fundamental ideas involved. From the measurement of wavelengths we can get valuable information regarding the structure of the molecules, the isotopic masses and the amount of work necessary to dissociate the atoms in it.

X-ray Spectra

(*i*) *The Line Spectrum.* As a consequence of the falling of the electrons from the outer to the inner shells we get *line* spectra characteristic of the element used for the target. However, unlike optical spectra which are of great complexity, the *X*-ray spectra of elements are all strikingly similar A typical pattern for an element of medium atomic weight The group of lines of shortest wavelength is known as *K*-series. Usually two lines of this series are detected (though there are a number of them). These lines are termed as K_α and K_β lines in the order of decreasing wavelengths. The K_α line is always the strongest in the spectrum. At longer wavelengths there is another group of lines at the *L*-series L_α L_β L_γ etc.), also in the order of decreasing wavelengths. The pattern is essentially the same for all elements. But as we ascend the periodic table the whole pattern shifts progressively towards shorter wavelengths with increasing atomic weight. For heavier elements a third series called the M-series has been detected ; while for the lightest elements even the *L*-series is beyond the range of our instruments and only the K-series can bc observed.

(*ii*) *The continuous spectrum.* In addition to the line spectrum, there exists a continuous X-ray spectrum. This spectrum is independent of the nature of the target except for the fact that the intensity is proportional to the atomic number of the element of which the target is made. The energy in the radiation is, however, of a minute fraction, about 1/1000 of that of the incident cathode rays, the remaining portion of the energy appearing as heat in the target.

The distribution of energy in the continuous background spectrum is found to depend only on the potential difference applied across the tube. The results obtained for different values of the potential difference. The maximum of the curve moves in the direction of shorter and shorter wavelengths as the potential difference across the tube is increased.

An important feature of these curves is that for any given potential difference used to excite the tube there is a sharply defined minimum wavelength of the radiation emitted by the tube or in other words, a maximum frequency which cannot be exceeded. According to classical electromagnetic theory in accelerating (or decelerating) charged particle should generate electromagnetic waves. It is, therefore, to be expected that radiation would be emitted by electrons brought suddenly to rest on striking the target of the X-ray tube.

However, the classical theory is quite unable to explain why there should be any *sharp cut-off* wavelength of the kind observed. Very careful investigation of this end part of the spectrum by Hall and others has shown that this frequency ν is directly proportional to the applied potential difference. Increasing V decreases λ_{min}. At a particular V, λ_{min} is the same for both tungsten and molybdenum targets, Duane and Hunt have found that λ_{min} (cut-off wavelength) is inversely proportional to V ; their precise relationship is

$$\lambda_{min} = \frac{1.24 \times 10^{-6}}{V} \text{ m}$$

where V is the *numerical value* of potential difference in volts.

This relation is called absorption limit for a particular series and is derived below.

Once again the quantum theory is applied in explaining this effect, according to which the emission or absorption of electromagnetic radiation takes place in packets of energy (photons) of amount given by

$$E = h\nu,$$

where h is Planck's constant and ν the frequency of waves.

If V is the potential difference in volts, it was found that over the whole range of potential differences employed from (20 to 100 kV) the relation could be expressed by the formula,

$$Ve = h\nu_{max}, \qquad \text{... (19)}$$

where e is the electronic charge and ν_{max}, the maximum X-ray frequency : Ve is the kinetic energy of the bombarding electrons which is being entirely spent in generating the photon. Thus the maximum frequency of radiation emitted is related to the maximum energy of a single cathode ray particle in the tube by the quantum relation.

The cut-off or minimum wavelength (λ_{min}) is given by

$$c = \nu_{max}\lambda_{min}, \qquad \text{... (20)}$$

where c is the velocity of light.

∴ combining Eqs. (19) and (20), we have

$$\lambda_{min} = \frac{c}{v_{max}} = \frac{hc}{h\nu_{max}} \qquad \text{... (21)}$$

$$= \frac{hc}{Ve} \qquad \text{... (21a)}$$

$$= \frac{6.625 \times 10^{-34}\ \text{J-s} \times 3 \times 10^{8}\ \text{ms}^{-1}}{1.6 \times 10^{-19}\ C \times V}$$

$$= \frac{1.24 \times 10^{-6}}{V} \text{ m.} \qquad \ldots [\because CV = J]$$

$$= \frac{12400 \times 10^{-10}}{V} \text{ m}$$

$$= \frac{12400}{V} \text{ Å} \quad \ldots [\because 1 \text{ Å} - 10^{-10} \text{ m}] \ldots (22)$$

Thus the quantum hypothesis predicts that the cut-off wavelength λ_{min} should vary inversely as the potential difference V applied across the tube ; and this is found to be in good agreement with the experimental results. Most of the electrons that generate X-ray photons give up only a part of their energy in this way and most of the X-radiation, therefore, is of longer wavelength than λ_{min}. Experiment shows that if a homogeneous beam of electrons falls on a very thin target the energy in the X-ray spectrum emitted is actually a maximum at the maximum frequency and diminishes as the inverse square of wavelength.

The cut-off wavelength λ_{min} can be measured with great accuracy with an X-ray spectrometer. Since c and e are already known very accurately, the equation may be used to find the Planck's constant h, and the result obtained by this method is found to agree very closely with that determined from photoelectric phenomenon. It may be noted that in the present case we are dealing with the transference of energy from an electron to radiation while in the photoelectric effect we have the *reverse* phenomenon but the same law governs both.

Moseley's law of characteristic X-ray spectra. In 1913 Moseley, an English Physicist investigated the relation between the characteristic X-ray spectra of various heavy elements and discovered that unlike optical spectra there existed a simple relationship between X-ray spectra and the order of the element in the Periodic Table. They are progressively shifted to higher frequencies and there is no trace of periodicity in them as in optical spectra. The frequency of any line, say, the first line of K_α series, is

$$\nu = a\,(Z - b)^2 \qquad \text{... (23)}$$

where Z is the atomic number of the element and a, b, are constants for that particular line. This relation is known as Moseley's law.

Thus from Moseley's law we find that the square root of the frequencies of the characteristics X-rays are related in a *linear* manner to the atomic number of the substances emitting the rays. Now let us compare Moseley's equation with the relation,

$$\nu = \frac{me^2}{8\varepsilon_0^2 h^3}\left(\frac{1}{n_1^2} - \frac{1}{n_2^2}\right)$$

for H spectrum of Eq. (15) which may be written as

$$\nu = \frac{Z^2 me^4}{8\varepsilon_0^2 h^3}\left(\frac{1}{n_1^2} - \frac{1}{n_2^2}\right) \qquad \text{... (24)}$$

$$= Z^2 Rc\left(\frac{1}{n_1^2} - \frac{1}{n_2^2}\right) \qquad \text{... (25)}$$

for an element with atomic number Z, where R is Rydberg constant and c the velocity of light.

Taking the simplest possible values $n_1 = 1$ and $n_2 = 2$

$$\nu = Z^2 Rc \times 3/4 \qquad \text{... (26)}$$

or

$$Z = \left(\frac{\nu}{3Rc/4}\right)^{1/2}$$

$$= Q\sqrt{\nu} \qquad \text{... (27)}$$

where

$$Q = \left(\frac{3}{4}Rc\right)^{-\frac{1}{2}} \qquad \text{... (28)}$$

It is clear from Moseley's curves between $\sqrt{\nu}$ and Z that if we take the K-radiation, Z increases by regular amounts as we pass up the series of elements in the periodic table and it was this fact that led him to the conclusion that the nuclear charge determines the characteristic radiation.

For $K_{\alpha 1}$ line of copper

$$\lambda = 1.54 \text{ Å}$$

$$= 1.54 \times 10^{-10} \text{ m}$$

$$\therefore \qquad \nu = \frac{c}{\lambda}$$

$$= \frac{3 \times 10^{8} \text{ ms}^{-1}}{1.54 \times 10^{-10} \text{ m}}$$

$$= 1.95 \times 10^{18} \text{ Hz.}$$

Also Rydberg's constant

$$R = 1.097 \times 10^{7} \text{ m}^{-1},$$

Z for copper = 29 and $e = 1.6 \times 10^{-19}$ C.

$\therefore$ by equation (26),

$$\nu = (29)^2 \times 1.097 \times 10^{7} \text{ m}^{-1} \times 3 \times 10^{8} \text{ ms}^{-1} \times \frac{3}{4}$$

$$= 2.07 \times 10^{18} \text{ Hz.}$$

This agreement is surprisingly good, considering the assumptions involved. Moseley showed that there is still a correction to be made for the effect of electrons upon each other. There is little doubt, therefore, that K-radiation originates when electrons jump directly from L–, M– or N-shell to the inner-most ring of the planetary electrons, which is called the K-ring or K-shell.

Bearing in mind the origin of the radiations, it appears that when an electron moving sufficiently rapidly through an atom removes one of the electrons from the *K*-shell, the return of an electron from outside probably the next shell gives rise to the characteristic radiation. The *K*-radiation thus corresponds in the case of heavier elements, to the Lyman series in the hydrogen spectrum.

Following the analogy, *L*-series will correspond to the Balmer series of hydrogen. So that if we put $n_1 = 2$ and $n_2 = 3$ in Eq. (25)

$$\nu = Z^2 Rc\left(\frac{1}{4} - \frac{1}{9}\right)$$

$$= Z^2 Rc \cdot \frac{5}{36} \qquad \text{... (29)}$$

Taking platinum $L_{\alpha 1}$ line

$$\lambda = 1.32 \times 10^{-10}\text{m},$$

$$\therefore \qquad \nu = \frac{c}{\lambda} = \frac{3 \times 10^8 \text{ ms}^{-1}}{1.32 \times 10^{-10}\text{ m}}$$

$$= 2.27 \times 10^{18} \text{ Hz}$$

and from Eq. (29) putting 78 for Z, we have

$$\nu = (78)^2 \times 1.097 \times 10^7 \text{ m}^{-1} \times 3 \times 10^8 \text{ ms}^{-1} \times \frac{5}{36}$$

$$= 2.78 \times 10^{18} \text{ Hz}$$

a fair agreement again.

Similarly, putting

$$n_1 = 3 \text{ and } n_2 = 4$$

$$\nu = Z^2 Rc\left(\frac{1}{9} - \frac{1}{16}\right) = Z^2 Rc\,\frac{7}{144} \qquad \text{... (30)}$$

Taking M_α line for thorium, $\lambda = 4.14 \times 10^{-10}$ m

$$\nu = \frac{c}{\lambda} = \frac{3 \times 10^8 \text{ ms}^{-1}}{4.14 \times 10^{-10} \text{m}}$$

$$= 0.725 \times 10^{18} \text{ Hz}$$

and since $Z = 90$ for thorium, above relation gives

$$\nu = (90)^2 \times 1.067 \times 10^7 \text{ m}^{-1} \times 3 \times 10^8 \text{ ms}^{-1} \times \frac{7}{144}$$

$$= 1.295 \times 10^{18} \text{ Hz.}$$

This discrepancy can be explained if we remember that for high atomic number elements this correction for the interaction of the electrons in the shells is very great. Apart from this the agreement is sufficiently near to recognise the analogy between the *M*-radiation and the Ritz-Paschen series of hydrogen.

In general, therefore, *K, L, M,* etc., series originate when electrons from cuter shells fall to these levels.

Moreover, from the equation of electrostatic attraction

$$mr\omega^2 = \frac{Ze \times e}{4\pi\varepsilon_0 r^2}$$

it follows that with increasing nuclear charge, $mr\omega^2$ increases so that as r diminishes with increasing attraction, ω the angular velocity which determines the frequency of revolution of the electron in their orbits, increases.

Ionization potential. We have already seen in the case of hydrogen that energy required to eject an electron from its orbit to infinity is the ionizing energy and this would correspond to the energy in the ground state, given by

$$E = -\frac{me^4}{4\pi\,\varepsilon_0^2\,h^2} \qquad \text{...}cf. \text{ Eq. } (12a)$$

$$= -\frac{9.11 \times 10^{-31}\ \text{kg} \times \left(1.6 \times 10^{-19}\text{C}\right)^4}{8\left(8.854 \times 10^{-12}\ \text{Fm}^{-1}\right)^2 \times \left(6.625 \times 10^{-34}\ \text{Js}\right)^2}$$

$$= -\frac{9.11 \times (1.6)^2 \times 10^{-15}}{8\,(8.854)^2 \times (6.625)^2}\ \text{J} \qquad \ldots [\because C^4 = F^2J^2]$$

$$= -21.76 \times 10^{-19}\ \text{joule.}$$

This is frequently represented in terms of the potential called the *ionization potential* and it measures the potential difference through which the electron must fall to acquire sufficient energy to detach itself from the atom or the molecule when the atom or the molecule is said to *be ionized*. This energy is generally measured in eV (electron-volts) and the corresponding voltage is the ionization potential. We have already seen in the case of hydrogen the good agreement between the calculated and the experimental values.

Ionization potential takes a special importance from the relation between X-rays and the electrons producing them or the electrons produced by X-rays. For the production of any given quality of X-rays the velocity of the cathode rays in the X-ray tube must have a certain minimum value. This minimum value, v_{min}, in the case of Eradiation was found by Whiddington to be $2(Z - 2)\ 10^6\ \text{ms}^{-1}$, where Z is the atomic number of the element of which the anti-cathode is constructed.

In the case of nickel, $Z = 28$, so that

$$v_{min} = 2\,(Z - 2) \times 10^6\ \text{ms}^{-1}$$

$$= 2 \times 26 \times 10^6\ \text{ms}^{-1}$$

$$= 5.2 \times 10^7\ \text{ms}^{-1}$$

The energy of the electron is, therefore,

$$\frac{1}{2}mv^2 = \frac{1}{2} \times 9.11 \times 10^{-21}\ \text{kg} \times \left(5.2 \times 10^7\ \text{m s}^{-1}\right)^2$$

$$= 1.23 \times 10^{-16}\ \text{joule.}$$

If this energy is converted into radiation and the quantum theory is applied to the process, then

$$1.23 \times 10^{-16}\ \text{J} = h\nu$$

$$\nu = \frac{1.23 \times 10^{-15}\ \text{J}}{6.625 \times 10^{-34}\ \text{J-s}}$$

$$= \frac{1.23 \times 10^{-10}\ \text{s}^{-1}}{6.625}$$

and

$$\lambda = \frac{c}{v} = \frac{3 \times 10^{8}\ \text{ms}^{-1}}{1.23 \times 10^{19}\ \text{s}^{-1}} \times 6.625$$

$$= 1.62 \times 10^{-10}\ \text{m}.$$

The observed value of λ for the two *K*-radiations of nickel are 1.66×10^{-10} m and 1.51×10^{-10} m, so that the agreement is sufficiently close to justify the application of quantum theory to this exchange. It is of course assumed that work function of the electron is negligible as compared to the energy of the bombarding electrons.

In an *X*-ray tube suppose that the energy of the incident electrons is so high that they are able to ionize even the inner-most shells of the atoms of the target. When this happens a vacancy is created in one of these shells, which can then be filled by a transition from a shell of higher energy. Because the inner-most shells are very tightly bound to the nucleus, the energy difference between them is considerable. The photon emitted in such a transition is, therefore, of high frequency and short wavelength—in fact in the *X*-ray region of the spectrum.

Energy levels. The detailed structure of the characteristic radiation, leaving out the fine structure can be reasonably understood from the energy level diagram. The energy level diagram for the different shells of an atom, the levels being indicated diagrammatically by horizontal lines. When there is a vacancy in the *K*-shell transitions are possible from all the outer shells *L, M, N, O* ..A transition from the *L* shell produces the K_α, line, that from *M*-shell gives the K_β line. The K_β transition involves a large change of energy and this line is, therefore, of shorter wavelength than the K_α line. However, this is a less probable transition and the K_β line is, therefore, weaker than the K_α line. Likewise, the *L*-series arises from transition following ionization of the *L*-shell. These are of lower energy than the *K*-series transitions and are, therefore, of much longer wavelengths. It is now clear why all *X*-ray spectra show the *same* pattern and also a *progression.*

The inner-most shell structure of all atoms is the same, since in all cases these shells are completely filled in the ground state. When electrons are ejected from the inner shells, the same pattern of transition is possible, whatever the nature of atoms forming the target. However, if one atom carries a larger charge on its nucleus than another its shells are more tightly bound and are, therefore, of lower energy. The corresponding changes of energy are then of larger value, giving a spectrum in which ail the lines have shorter wavelengths. It, therefore, seemed that the atomic number *Z* which Moseley showed to be of such significance in fixing the properties of the atom, was to be identified with the total charge on the nucleus. It is equal to the number of planetary electrons in the neutral atom.

The explanation given above of the main lines of an *X*-ray spectrum does not account for the fine structure in various series because we have taken into consideration only the *principal quantum numbers* of the different shells. If other quantum numbers are considered then every energy shell except *K* has a number of sub-shells, with energies slightly differing from each other so that transitions are possible between these sub-shells of any particular main shell and the lower shell. There are three sub-shells in *L* level, 5 in *M* level and 7 in *N* level and the possible transitions give rise to $K_{\alpha1}$, $K_{\alpha2}$, $K_{\beta1}$, $K_{\beta2}$, $L_{\alpha1}$; $L_{\beta1}$, $L_{\gamma1}$, $L_{\gamma2}$ etc.

The Raman effect. The effect deals with the intercharge of energy between radiation and matter during the scattering of light. When a beam of light is passed through a transparent medium solid, liquid or gaseous it is found that however clear and homogeneous the medium may be, there is always a certain amount of light scattered sideways out of the beam. Tyndall and Rayleigh had first studied the phenomenon of scattering of light by molecules by passing a strong beam of light through a pure liquid and the latter had explained that the blue colour of the light scattered laterally from the beam was due to the fact that *the scattering was inversely proportional to the fourth power of the wavelength,* that is,

$$S \propto \frac{1}{\lambda^4},$$

where S denotes the amount of scattering and λ the wavelength of the light scattered.

In these experiments it was assumed that the scattered light consisted of some of the components of the incident beam. In 1928, however, Sir C.V. Raman after several years of investigation of light scattering, discovered that when a beam of monochromatic light of frequency v is made to traverse a clear homogeneous and transparent medium —solid, liquid or gaseous and the scattered light is spectrographed with a high resolving power instrument, several lines make their appearance on *either* side of the line of incident frequency having frequencies $\nu \pm \nu_1, \nu \pm \nu_2, \nu \pm \nu_3 \cdots$

It is found that the frequencies $\nu_1, \nu_3, \nu_3 \ldots\ldots$ are quite *independent* of the frequency v of the incident beam and are *characteristic* of the molecules of the scattering substance. These lines of modified wavelengths arc called Raman Lines and their appearance is known as Raman effect. They are of course very much weaker than the unmodified lines, those of *smaller* frequency being more *intense* than those of *higher* frequency.

Raman effect is somewhat similar to fluorescence but it differs from it in two important respects. *Firstly,* the light which is incident on the scattering substance must have a wavelength which does not correspond to one of the absorption lines or bands of the substance otherwise fluorescence will result, for in this case the rediated energy is of a frequency lower than that which the fluorescent substance is able to absorb. *Secondly,* the intensity of light in Raman scattering is much less than most fluorescent light.

Study of Raman effect. Raman's experimental arrangement. Light from a mercury arc M is focused by a condenser L_1 on to the liquid or gas under examination contained in a non-fluorescent glass tube A. The light scattered in a direction at right angles to the incident beam is collected by a lens L_2 and focused on the slit S of the spectrograph. The slit is suitably shielded from all stray light by means of screens PP. To obtain monochromatic light of greater intensity saturated solution of sodium nitrate is placed in the path of the mercury arc, since it will absorb the ultra-violet lines of the arc and transmit the blue-violet line only. An exposure of several hours reveals lines on *either* side of the incident line. These lines are Raman lines.

The effect has been studied in the case of a large number of substances—liquids and gases and the results have provided a great deal of information about the molecular structure.

The theory of the band spectra of molecules sketched above was also successfully verified by Sir C.V. Raman in his experiments on the molecular scattering of light.

Explanation of Raman spectrum. A satisfactory explanation of the Raman spectrum has been given on the quantum theory of radiation. Smekal working in America had in fact anticipated the results earlier on the basis of the quantum theory.

When a photon of light is incident on a molecule

(*i*) it may be scattered or deflected off its path without loss or gain of energy thus giving rise to unmodified lines as referred to earlier in Tyndall scattering, or

(*ii*) during the collision between the photon and the molecule some of the energy of the photon may be used in raising the molecule from a lower to a higher value of the vibrational or rotational energy level and the photon may thus be scattered with a lower energy $h\nu' = (h\nu - h\nu_1)$, so that the scattered light has a lower frequency $\nu' = (\nu - \nu_1)$, or

(*iii*) it may happen that the photon collides with a molecule which is already in an excited state and is about to return to the *normal* state or its arrival may precipitate the return, the energy thus set free attaches itself to the striking photon which, therefore, flits off with gain of energy giving rise to scattered light of higher frequency $(\nu + \nu_1)$, and hence smaller wavelength. The difference between the frequency of each Raman line and the frequency of the incident light is thus equal to the frequency of some conceivable emission or aborption line of the scattering atom or molecule. Moreover since the number of molecules in the excited states would be smaller compared to those in the normal state, the latter process is less likely to occur than the former and it is therefore easily understandable that the Ran an lines on the shorter wavelength side would be weaker than their counterparts on the longer wavelength side. The lines displaced to the red end are referred to as *Stoke's lines* and those to the violet end as *anti-Stoke's lines*. This is due to the Stoke's hypothesis proposed some years earlier that the secondary radiation such as fluorescence was always of longer wavelength than incident light. It may be interesting to note that with the increase of temperature the intensity of the high frequency lines increases. This is due to the fact that the number of excited molecules would increase with increase of temperature.

Applications of Raman Spectrum

(1) The relative intensities of the Raman lines give information about the molecular structure of compounds particularly regarding their rotational and vibrational frequencies even though we might be unable to determine these by absorption or emission spectra. As explained above the small frequencies of the molecular spectrum become the differences of frequencies in the visible spectrum.

(2) It enables us to study the internal structure of liquids, gases and solids, the arrangement of atoms in a particular molecule and electronic structure of elements. It enables the energy of levels, which would not otherwise be accessible, to be measured.

(3) In industry Raman spectrography is being applied to study the properties of plastics.

(4) The effect is successfully used to study the compositions of a mixture and particularly in the analysis of aromatic oils.

(5) Recently Raman effect has been used for the diagnostic study of plasma, that is, for determining the importance of certain plasma processes, such as recombination rates, diffusion constants, energy transfer processes and chemical activity in the plasma. By measuring the relative intensity of Stokes and anti-Stokes Raman lines a determination can be made of the density and temperature of individual species present in a plasma.

The Compton effect and photo-momentum. The Compton effect demonstrates in a remarkable manner the validity of the quantum theory of radiations as well as some of the predictions of the theory of *relativity*. According to photoelectric effect radiation reveals a corpuscular nature with each quantum of energy $h\nu$ when it is transferred and converted into the energy of electrons, but Compton effect goes a step further and establishes that radiation preserves its quantum character even when it is freely propagated in space and further that the quanta of energy or photons, as they are called have a mass $h\nu/c^2$ in accordance with the principle of relativity and carry a definite photo-momentum $\frac{h\nu}{c}\left(=\frac{h}{\lambda}\right)$ shown below :

On quantum theory the energy of a photon is,

$$E = h\nu$$

Einstein's relativity equation for the equivalence of mass and energy gives $E = mc^2$, where m is the mass and c the velocity of light. Therefore a photon of energy hv has a mass given by hv/c^2. Since the momentum of a particle is the product of its mass and velocity, the momentum of the photon becomes,

$$P = mc = \frac{h\nu}{c^2} \times c$$

$$= \frac{h\nu}{c} = \frac{h}{\lambda} \quad \text{... (31)}$$

If this photon strikes a free electron at rest the collision may be treated exactly in the same way as the collision between two perfectly elastic particles. If we assume that the principles of 'Conservation of Energy' and 'Conservation of Momentum' hold during this process, then as a result of this collision, the electron will acquire a velocity v in a direction making an angle θ with the direction of motion of the incident photon and a photon of reduced frequency ν_1 will be scattered in the direction ϕ with the original direction.

To obtain a solution of the impact we must express the kinetic energy and the momentum of the electron not in their usual form but in the form given by *Relativity Principle.* The kinetic energy of the electron then becomes

$$E_{kin} = (m - m_0)c^2 = \left[\frac{m_0}{\left(1-\beta^2\right)^{\frac{1}{2}}} - m_0\right]c^2$$

... *cf.* Eqn. (25-26)

$$= m_0 c^2 \left[\frac{1}{\left(1-\beta^2\right)^{\frac{1}{2}}} - 1\right] \quad \text{... (32)}$$

and its momentum

$$mv = \frac{m_0 \beta c}{\left(1-\beta^2\right)^{\frac{1}{2}}} \quad \text{... } cf \text{ Eqn. (25-23) ... (33)}$$

where m is the relativistic mass of the electron, m_0 its rest mass and

$$\beta = \frac{v}{c}.$$

By the principle of 'Conservation of Energy' we have

$$h\nu = h\nu_1 + m_0 c^2 \left[\frac{1}{\left(1-\beta^2\right)^{\frac{1}{2}}} - 1\right] \quad \text{... (34)}$$

Resolving the momentum vectors into two rectangular components along and at right angles to the direction of the incident photon and using the principle of conservation of momentum, we get

$$\frac{h\nu}{c} = \frac{h\nu_1}{c}\cos\phi + \frac{m_0 \beta c}{\left(1-\beta^2\right)^{\frac{1}{2}}}\cos\theta \quad \text{... (35)}$$

$$0 = \frac{hv_1}{c}\sin\phi + \frac{m_0\beta c}{\left(1-\beta^2\right)^{\frac{1}{2}}}\sin\theta . \qquad \dots (36)$$

Since we want to find shift in wavelength, let us introduce wavelengths $\lambda = \frac{c}{v}$ and $\lambda_1 = \frac{c}{v_1}$ of the incident and the scattered radiation respectively. The Equations (35) and (36), then become

$$\frac{h}{\lambda} - \frac{h}{\lambda_1}\cos\phi = \frac{m_0\beta c}{\left(1-\beta^2\right)^{\frac{1}{2}}}\cos\theta$$

$$\frac{h}{\lambda_1}\sin\phi = \frac{m_0\beta c}{\left(1-\beta^2\right)^{\frac{1}{2}}}\sin\theta .$$

Squaring these two equations and adding, we get

$$\frac{h^2}{\lambda^2} + \frac{h^2}{\lambda_1^2} - \frac{2h^2\cos\phi}{\lambda\lambda_1} = \frac{m_0^2\,\beta^2 c^2}{1-\beta^2} \qquad \dots (37)$$

Similarly, equation (34) can be written as

$$\left(\frac{h}{\lambda} - \frac{h}{\lambda_1}\right) + m_0 c = \frac{m_0 c}{\left(1-\beta^2\right)^{\frac{1}{2}}}$$

On squaring it' we get

$$\left(\frac{h^2}{\lambda^2} - \frac{h^2}{\lambda_1^2} - \frac{2h^2}{\lambda\lambda_1}\right) + 2m_0ch\left(\frac{1}{\lambda} - \frac{1}{\lambda_1}\right) + m_0^2c^2 = \frac{m_0^2c^2}{1-\beta^2}$$

or $$\frac{h^2}{\lambda^2} + \frac{h^2}{\lambda_1^2} - \frac{2h^2}{\lambda\lambda_1} + 2m_0ch\left(\frac{1}{\lambda} - \frac{1}{\lambda_1}\right) = \frac{m_0^2\,\beta^2 c^2}{1-\beta^2} \qquad \dots (38)$$

Subtracting equation (37) from (38), we get

$$\frac{2h^2}{\lambda\lambda_1}(\cos\phi - 1) + 2m_0 ch\left(\frac{1}{\lambda} - \frac{1}{\lambda_1}\right) = 0.$$

Therefore the change in wavelength

$$\Delta\lambda = \lambda_1 - \lambda = \frac{h}{m_0 c}(1 - \cos\phi) \quad \text{... (39)}$$

$$= \frac{2h}{m_0 c}\sin^2(\phi/2) \quad \text{... (39a)}$$

If values $h = 6.625 \times 10^{-34}$ J-s;

$m_0 = 9.11 \times 10^{-31}$ kg;

and $c = 3 \times 10^8$ ms^{-1}

are inserted, then

$$\Delta\lambda = \lambda_1 - \lambda = 0.0484 \times 10^{-10} \sin^2(\phi/2) \text{ m}$$

$$= 0.0484 \sin^2(\phi/2) \text{ Å} \quad \text{... (40)}$$

which for radiation scattered at right angles to the incident beam gives us

$$\Delta\lambda = 0.0242 \text{ Å}, \quad \text{... (41)}$$

a value identical with that found by experiment, $\Delta\lambda = 0.0236$ Å.

Equation (39) states that the wavelength of the ray scattered at an angle ϕ should always be greater than the wavelength of the incident radiation. Further more, this difference in wavelength should not depend upon the nature of the scattering material but should depend only on the angle of scattering.

The factor $h/m_0 c$ with m_0 the rest mass of the electron is called the Compton wavelength of the electron and is usually designated by the symbol λ_c which on setting the values of h, m_0 and c gives

$$\lambda_c = 0.0242 \text{ Å}. \quad \text{... (42)}$$

The kinetic energy of the recoil electron can be shown to be,

$$E_{kin} = \frac{h\nu\left(1-\cos\phi\right)\alpha}{1+\alpha\left(1-\cos\phi\right)}, \quad \text{... (43)}$$

where $$\alpha = \frac{h\nu}{m_0c^2}.$$

If instead of being free to move the electron with which the photon collides is tightly bound to the parent atom, m_0 becomes the mass of the whole atom instead that of the single electron and $d\lambda$ thus becomes vanishingly small. The photon is then scattered without loss of energy and hence without change in wavelength.

On Compton Theory this accounts for the unmodified radiation. From equation (39) we find that the change in wavelength $d\lambda$ will be greatest when $\phi = 180°$; the magnitude of $d\lambda$ being $2h/m_0c$, is twice the Compton wavelength. This shows that the change in wavelength of the electromagnetic radiation interacting with the electron will never exceed $2h/m_0c$. Such small changes are readily observable in *X*-ray spectrum because of its larger relative percentage of the incident wavelength which is of the order of 1 Å. These changes in wavelength have been measured. A beam of monochromatic *X*-rays of known wavelength is directed at a target.

The scattered rays received at different scattering angles contain both the unmodified component and the altered wavelength. In this case the line of the molybdenum was scattered by graphite and measurements were made at scattering angles of 45°, 90° and 135°. Each scattered beam is seen to consist of two distinct lines. One line *P* has a wavelength corresponding to the wavelength of the incident radiation. The second line *M*, a bit apart has a longer wavelength λ, which depends upon the angle of scattering The wavelength λ_1 of this *modified* line is in good agreement with that calculated from equation (39). Similar results were obtained when other substances were used as scatterers. In all cases the wavelength of the modified line was found to depend only upon the angle of scattering and not upon the nature of the scattering substance as predicted by the theory. The Compton effect, therefore, is the most vivid and convincing evidence in favour of the view that light is corpuscular in nature.

Example. *In a X-ray scattering experiment the Compton shift was found to be 0.0242 Å for a scattering at 90°. Find the electronic charge.* ($e/m = 1.76 \times 10^{11}$ *C* kg^{-1}, $h = 6.624 \times 10^{-34}$ *J-s and* $c = 3 \times 10^8$ ms^{-1}).

$$\Delta\lambda = \frac{2h}{mc} \cdot \sin^2 (\phi / 2)$$

Substituting the values, we get

$$0.0242 \times 10^{-10} \text{ m} = \frac{2 \times 6.624 \times 10^{-34} \text{ J-s}}{m \times 3 \times 10^8 \text{ ms}^{-1}} \cdot \sin^2 45°$$

whence
$$m = \frac{2 \times 6.624 \times 10^{-34}}{3 \times 10^8 \times 0.0242 \times 10^{-10}} \times \left(\frac{1}{\sqrt{2}}\right)^2 \text{ kg}$$

Now
$$e/m = 1.76 \times 1011 \ C \text{ kg}^{-1}$$

Substituting for m, we get

$$e = \frac{6.624 \times 10^{-34} \times 1.76 \times 10^{11}}{3 \times 10^8 \times 0.0242 \times 10^{-10}}$$

$$= \mathbf{1.6058 \times 10^{-10}} \ C$$

Diffraction of X-rays by crystals—Bragg's law. It was suggested by Laue in 1913 that a crystal could be employed for the diffraction of X-rays. It was because the wavelength of X-rays was of the same order (10^{-8} cm) as the inter-atomic distance in a crystal. Soon afterwards Sir William Bragg devised a simpler way of analysing the diffraction of X-rays by a crystal lattice and succeeded in diffracting X-rays from a sodium chloride crystal. According to him crystals could be employed both as transmission and reflection gratings and the scattering of *X*-rays by crystals could be considered as reflection from successive atomic planes in the crystal.

Let us consider parallel lattice planes (equidistant from one another) in the crystal structure separated by a distance '*d*' Suppose a beam of *X*-rays is incident on a layer of atoms in one plane of the crystal at a glancing angle θ. Each atom in each plane scatters a part of *X*-ray beam in all directions, *i.e.*, it becomes a source of small spherical wavelets of X-rays. If in a certain direction the scattered *X*-rays from all the atoms in a given plane are in the same phase, they will reinforce one another and also if the composite waves from the successive atomic planes reinforce, the reflected beam will be quite intense. These directions are those for which the path difference for the waves reflected from the adjacent layers of atoms is a whole number of wavelengths. This path difference is equal to $PQ + QR$. For strong reflection,

$$PQ + QR = n\lambda$$

or

$$2\, d \sin\theta = n\lambda, \quad \text{... (44)}$$

where λ is the wavelength and n is an integer.

This is known as *Eragg's law* and is the basis for the determination of crystal structure. For any given wavelength λ and spacing d between the parallel crystal planes we can expect to find several angles at which maximum reflection occurs corresponding to $n = 1, 2, 3$, etc. Thus knowing λ and θ, d may be evaluated. By rotating the crystal to produce reflection from different sets of planes, the shape and value of the unit cell can be derived. If the density of the crystal is also known, then from the volume of the unit cell, it is possible to evaluate the *number* of atoms of various kinds in the cell.

Bragg's analysis was fully confirmed by experiment. It provided the means of measuring not only the inter-atomic spaces in crystals but also the wavelengths of X-rays emitted by any given source. Generally in experiments on X-rays n is set equal to unity. The X-rays diffraction techniques employed in the study of crystals are of two types :

(*i*) Rotating crystal technique and

(*ii*) powder technique.

For this purpose Bragg developed an instrument which is called *Bragg's X-ray spectrometer.*

Bragg's X-ray spectrometer. A mono-chromatic and narrow X-ray beam from an X-ray tube is obtained by the slits in the two lead screens S_1 and S_2 and is allowed to strike the crystal C mounted on a turn-table capable of rotating about a vertical axis passing through O. It carries a graduated scale. After reflection from the crystal surface the reflected beam is pawed through another slit S_3 and then into the ionization chamber I with lead walls.

This is connected to a suitable source of potential difference and an electrometer to measure the ionization current, which is taken as a measure of the strength of the reflected beam. With the help of turn-table the crystal is gradually rotated so as to vary the *glancing angle* θ : the angle between the ionization chamber and the primary beam is kept equal to 2θ in order to receive the beam reflected from the crystal and the ionization is measured as a function of θ. The results of such an experiment with a crystal of sodium chloride and the glancing angles for which the intensity of the reflected beam is maximum are given by the Bragg's equation,

$$2d \sin \theta = n\lambda.$$

The spectrum may consist of several orders. The maxima A_1 and B_1 correspond to two characteristic radiations in the first order spectrum of incident beam. The maxima A_2, B_2 and A_3, B_3 are produced by second and third order spectrum. Such a curve is called an *X-ray spectrum*. The smallest angle at which the reflection is maximum corresponds to $n = 1$. This is known as the first order reflection or spectrum : if n =2, 3, a second order, a third order spectrum is obtained, and so on. It is found that the maxima A_1 A_2 and A_3 occur at angles which are nearly 11.8°, 23.5° and 35.0°, such that

$$\sin 11.8° : \sin 23.5° : \sin 36.0° :: 0.204 : 0.400 : 0.600 :: 1 : 2 : 3 \text{ nearly.}$$

From the X-ray spectra obtained with different crystals keeping the target from which X-rays are emitted the same, we find that the same lines appear with about same relative intensity, but the angles at which they appear are changed, showing that the grating element d between the layers of atoms is different for different crystals. Conversely, having evaluated spacings for a crystal in a particular plane, this crystal could now be used for the measurement of wavelength of any other X-radiation.

Example. *Calculate the wavelength of X-rays from the following data :*

(*i*) *Glancing angle for NaCl for the first order spectrum* = 11.8°

(*ii*) *Atomic weight of Na* = 23

(*iii*) *Atomic weight of chlorine* = 35.5

(*iv*) *Density of NaCl* = 2.17×10^3 kg-m^{-3}

(*v*) *Mass of H_2 atom* = 1.64×10^{-24} g = 1.64×10^{-27} kg.

On examining the structure of NaCl crystal we find that each atom lies at the junction of eight cubes and that each cube has 8 atoms one at each corner, so that in the whole structure extending in all directions there is one atom per cube or half a molecule per cube.

∴ average mass of each cube

$$= \tfrac{1}{2} \text{ molecular weight} \times \text{weight of } H_2 \text{ atom}$$

Now molecular weight of NaCl

$$= 23 + 35.5 = 58.5$$

Mass of H_2 atom

$$= 1.64 \times 10^{-27} \text{ kg}$$

$\therefore$ average mass of each cube

$$= ½ \times 58.5 \times 1.64 \times 10^{-27} \text{ kg.}$$

But volume of each cube

$$d^3 = \frac{\text{mass}}{\text{density}}$$

$$= \frac{\frac{1}{2} \times 58.5 \times 1.64 \times 10^{-27} \text{ kg}}{2.17 \times 10^3 \text{ kg-m}^{-3}}$$

where d is the side of the cube,

$$\text{whence } d = \sqrt[3]{\frac{58.5 \times 1.64 \times 10^{-30} \text{ cm}^3}{2 \times 2.17}}$$

$$= \mathbf{2.81 \times 10^{-10} \text{ m}}$$

Now the glancing angle θ_1 for the first order spectrum of NaCl = 11.8°.

$$\lambda = 2d \sin \theta_1$$

$$= 2d \sin 11.8°$$

$$= 2 \times 2.81 \times 10^{-10} \times 0.204$$

$$= \mathbf{1.15 \times 10^{-10} \text{ m} = 1.15Å}$$

Example. *A neutron beam of K.E. 0.04 ev is diffracted at the (100) plane of the sylvine crystal (d_{100} = 3.14 Å). Calculate the glancing angle θ at which the first order Bragg spectrum will be observed. Mass of neutron =1.6748 × 10^{-27} kg and Planck's constant, h = 6.625 × 10^{-34} J-s.*

$$K.E.\left(\frac{1}{2}mv^2\right) = 0.04 \times 1.6 \times 10^{-19}\text{ J}$$

...[$\because$ 1 ev = 1.6×10^{-10} Joules]

$$\therefore \qquad v = \sqrt{\frac{2 \times 0.04 \times 1.6 \times 10^{-19}}{m}}$$

and
$$mv = \sqrt{2 \times 0.04 \times 1.6 \times 10^{-19}\,\text{m}}$$

$$= \sqrt{0.128 \times 10^{-19}\,\text{J} \times 1.6748 \times 10^{-27}\,\text{kg}}$$

$$= 10^{-27}\sqrt{1280 \times 16748\ \text{kg - ms}^{-1}}$$

Now de Broglie wavelength

$$\lambda = \frac{h}{mv}$$

Substituting for h and mv, we have

$$\lambda = \frac{6.625 \times 10^{-24}\,\text{J.s}}{10^{-27}\sqrt{1280 \times 16748\ \text{kg ms}^{-1}}}$$

$$= \frac{6.625 \times 10^{-10}}{4630}$$

$$= \mathbf{1.431 \times 10^{-10}\ m.}$$

From Bragg's law

$$2d \sin \theta_1 = \lambda.$$

$$\therefore \qquad \sin \theta_1 = \frac{\lambda}{2d} = \frac{1.431 \times 10^{-10}}{2 \times 3.14 \times 10^{-10}}$$

$$= 0.2278$$

$$\therefore \qquad \theta = \mathbf{13^\circ\ 10'.}$$

Particle and wave aspects of radiation. The photoelectric effect demonstrated that light propagates as a series of small discrete packets of energy. They are often referred to as photons. This is the *particle aspect* of radiation. At the same time we do not deny the *wave aspect* of the light, since in expressing the energy of a photon we always associate it with the *frequency* of the wave which is still regarded to be the electro-magnetic type described by Maxwell's theory. We, therefore, are led to the conclusion that radiation has *dual* aspect so that we have sometimes to regard it as particle and sometimes as a wave.

At first sight we are unable to reconcile to this view on account of our preconceived notions about particle and waves. A particle has a definite *position* in space and possesses *mass, velocity, momentum* and *energy* while a wave is associated with *frequency, wavelength, amplitude* and *intensity,* etc. When we talk of a particle, we think of billiard balls, bullets etc., which keep their concentrated entities and if waves are mentioned we think of them as spreading out from a source in the same way as ripples spread out on the surface of water in a pond, when a stone is thrown into it.

Moreover, according to the wave aspect, the energy is distributed continuously throughout the wave-pattern and if we take a motion" picture of the ripples on the surface of a pond we could thereby give an exact picture of the progress of the wavefront. The question is : "Could we follow the path of a photon in the same way ?" Obviously not, we shall explain it by an illustration.

Consider a point source of light on the axis of a convex lens, which forms a real image of the source on the side of it, say on a screen. There is no doubt that a photon emitted by the source will fall on the screen in the image spot. But the question is : Could we locate the position and trace the path of the photon in the intervening stage ? This is rather difficult to answer. In order to know whether the photon passed through a given point let us put a detector say, a photo cell there. But the cell will absorb and destroy the photon, so that the subsequent journey is not completed at all and we have no means of tracing its path.

But in spite of the difficulties, the photon seems to have its identity as the quantum theory is marvelously successful in explaining the photoelectric phenomenon. It predicts correctly that the maximum energy of the photoelectrons should depend upon the frequency of incident light and not upon its intensity contrary to what the wave theory suggests. The wave theory also cannot account for the threshold frequency as does the quantum theory, nor it can explain why even the feeblest light can lead to the instantaneous photo-emission. On the other hand, we cannot lose sight of the fact that the wave theory has very successfully explained a host of phenomena, notably interference, diffraction and polarization.

So we have now two theories of the nature of light each explaining a certain class of phenomena apparently inexplicable by the other. However, it may be noted that in a specific event light exhibits either a wave or particle aspect, never both simultaneously. The same light beam which produces interference and diffracts at a grating surface can cause photo-emission from a suitable metal surface but these processes occur independently. We can, therefore, conveniently say that *wave theory of light and the quantum theory of light complement each other.*

Einstein, Bohr, Schroedinger, Heisenberg, Dirac and others have made wonderful contributions towards the solution of this mystery. A deeper understanding of some of the excellent experiments suggests, that both theories are essentially correct as well as they go. It is hoped that in the not too distant future the subject of *Wave mechanics* discovered by Schroedinger in 1926 which is the name given to the new method of interpreting physical phenomenon may provide the fundamental principle necessary for reconciling the apparently conflicting ideas. According to wave mechanics one interpretation of the wave particle description of light is to regard the *intensity of the wave* at a given point as a measure of the *probability* of the photon appearing there. The path of the photon cannot be predicted with certainty but the probability of its following a particular path can be stated with certainty. Whether or not it actually goes that way is only to be stated *after* the event. This variable quantity *probability* is called the *wave function.* Waves of probability and particles that can only be detected as said above by making them vanish are certainly very tough concepts which need further clarification.

5

Electromagnetic Theory

Upto Fresnel's time light was studied as an isolated subject but after the discovery of *Faraday effect* in 1845 regarding the rotation of plane of polarisation of a beam of light by certain isotropic substances like flint glass when placed in a powerful magnetic field, it was inferred that there was some close connection between light and magnetism. As a consequence of this discovery, Faraday proposed that ultimate atom is nothing more than a field of force surrounding a point centre.

The electromagnetic waves. It was in an intuitive attempt by Maxwell, a British Physicist (1831—1879), to explain the properties of a medium that would transmit electric action that the foundations of electromagnetic theory of light (classical Field theory as we call it) were laid. Maxwell having studied the work of Faraday thoroughly accepted his concept of lines of force to visualize strain in an ether surrounding the electrified body. He farther introduced the revolutionary idea of *displacement currents* in dielectrics. According to this if the plates of a condenser C, are joined to the terminals of a battery

B, during the short initial time in which the condenser plates are being charged and Faraday's tubes are being established, there exists in the dielectric between the plates a momentary current which flows as long as the electric field *E* is changing and stops as soon as *E* becomes constant.

The magnitude of this momentary current is proportional to the rate at which the number of Faraday's tubes is increasing and it produces the same magnetic effects, of course momentary like itself, as an ordinary conduction current. We also know that induced e.m.f. results whenever there is a change in the magnetic field, it obviously follows that a momentary magnetic field must be associated with electric intensity.

This electric intensity will give rise to another displacement current which in turn will produce another momentary magnetic field and so on. It follows, therefore, that if there is an initial displacement current as postulated by Maxwell, a state of periodic variations of the electric and magnetic fields will be propagated in the medium. This constitutes an *electromagnetic* disturbance of a very high frequency given by

$$f = \frac{1}{2\pi\sqrt{LC}} \qquad \text{... (1)}$$

where L and C are respectively the inductance and capacitance of the circuit.

According to Maxwell the displacement currents were not restricted only to dielectrics as supposed by Faraday, but they occurred whenever there was a changing electric field. Using this generalised idea of displacement currents. Maxwell expressed the results of his theoretical investigations in the form of a set of differential equations describing the electromagnetic field known as Maxwell's equations which are expressed in the concise differential forms in vector notation as follows :

Maxwell's Equations of Electromagnetic Field

$$\left.\begin{aligned} \nabla \times \boldsymbol{E} &= -\frac{1}{c}\cdot\frac{\partial \mathbf{B}}{\partial t} && \text{... }(a) \\ \nabla \times \mathbf{B} &= -\frac{4\pi J}{c}+\frac{1}{c}\cdot\frac{\partial E}{\partial t} && \text{... }(b) \\ \nabla \,.\, \mathbf{E} &= 4\pi\rho && \text{... }(c) \\ \nabla \,.\, \mathbf{B} &= 0 && \text{... }(d) \end{aligned}\right\} \quad \text{...(2)}$$

where *E, B* and *J* are the vectors which govern the electromagnetic field and are called as :

E = Electric field intensity in e.s.u.

B = Magnetic field intensity in gauss.

J = Conduction current density in e.s.u./sec.

π = Charge per unit volume.

These equations are in fact the differential forms of important electrical laws.

Equation (2*a*) expresses *Faraday's law of magnetic induction.*

Equation (2*b*) expresses the dependence of magnetic field on conduction current density *J* and displacement current density

$$J_d = \left(=\frac{1}{4}\cdot\frac{\partial E}{\partial t}\right), \textit{viz., Ampere's circuital law.}$$

Equation (2*c*) is the statement of *Coulomb's law* of flux of induction from a charge ; and

Equation (2*d*) expresses that magnetic field is due to currents alone.

In space devoid of any matter, whatsoever (vacuum), the terms **J** and **P** are zero and the Maxwell's Equations reduce to:

$$\left.\begin{aligned} \nabla \times \mathbf{E} &= -\frac{1}{c}\cdot\frac{\partial B}{\partial t} && \text{...}(a) \\ \nabla \times \mathbf{B} &= \frac{1}{c}\cdot\frac{\partial E}{\partial t} && \text{...}(b) \\ \nabla \,.\, \mathbf{E} &= 0 && \text{...}(c) \\ \nabla \,.\, \mathbf{B} &= 0 && \text{...}(d) \end{aligned}\right\} \quad \text{...(3)}$$

Here the displacement current $\left(\frac{1}{c}\cdot\frac{\partial E}{\partial t}\right)$ plays the all important part because if there is a displacement current as postulated by Maxwell a state of periodic variations of the intensities of electric and magnetic fields will be propagated in the space.

Derivation of Maxwell's Equations

(*i*) Equation (3*a*) is the expression of *Faraday's law of magnetic induction,* as said above, which may be stated as follows :

If C is a closed curve, S the surface enclosed by it, and ϕ the flux through it by, say, suddenly starting an electric current in a coil of wire in its neighbourhood, then there is set up an induced e.m.f. e around the closed path given by

$$e = \int_C \mathbf{E}.\mathrm{ds} = -\frac{1}{\mathrm{c}}.\frac{\mathrm{d}}{\mathrm{dt}}\int_S \mathbf{B}.\mathrm{da}\ .\ da = \frac{1}{c}.\frac{d\Phi}{dt} \quad \text{... (4)}$$

where **E** is the electric intensity round the closed path in e.s.u.

B is the magnetic field in Gauss, ϕ is the total flux of induction in Gauss cm^2, and $e = \int_C \mathbf{E}\,\mathrm{ds}$ in statvolts, c is the ratio of 1 e.s.u. pot. diff. to 1 e.m.u. of pot. diff. (numerically equal to the velocity of e.m. waves in vacuum = 3×10^{10} cm/

sec., the negative sign is taken because the electromotive force produced always opposes the current producing it.

Using the vector derivative curl and Stokes' theorem Eq. (23) of Appendix V *viz.*,

$$\int_C \mathbf{E}\, ds = \int_S \text{curl } \mathbf{F}.\, da.$$

We can express this law as :

$$\text{Curl } \mathbf{E} = \frac{1}{c}.\frac{d\mathbf{B}}{dt} \qquad \text{... (5)}$$

assuming that Eq. (5) holds for an infinitesimal area.

Recognizing that **B** may depend upon position as well as time we shall write — for $\frac{\partial \mathbf{B}}{\partial t}$ for $\frac{d\mathbf{B}}{dt}$. We have then these two entirely equivalent statements of the law of induction.

$$\left.\begin{aligned} \int_C \mathbf{E}\, ds &= -\frac{1}{c}.\frac{d}{dt}\int_S \mathbf{B}.\, da \\ \text{or} \qquad \text{Curl } \mathbf{E} &= -\frac{1}{c}.\frac{\partial \mathbf{B}}{\partial t} \\ \textit{i.e.,} \qquad \nabla \times \mathbf{E} &= -\frac{1}{c}.\frac{\partial \mathbf{B}}{\partial t} \end{aligned}\right\} \qquad \text{... (6)}$$

(*ii*) Eq. (3*b*) indicates the dependence of magnetic field on conduction and displacement currents. According to *Ampere's circuital law* the line integral of magnetic field B around a closed path C surrounding a wire carrying a current *I* is given by

$$\int_C \mathbf{B}\cdot ds = \frac{4\pi I}{c} \qquad \text{... (7)}$$

where *I* is in e.s.u. *f* sec, *B* in Gauss, s in cm and $c = 3 \times 10^{10}$ cm/sec.

or $$\int_C \mathbf{B} . ds = \frac{4\pi}{c} \int_S \mathbf{J} . da \qquad ... (8)$$

where J is conduction current density *i.e.*, current per unit area of the surface S, enclosed by the curve. Comparing Eq. (8) with Stocks' theorem as in deriving equation (3*a*) we find that a statement equivalent to Eq. (8) is as follows :

$$\text{curl } B = 4\pi J/c \qquad ... (9)$$

Eq. (9) is correct for a steady closed circuit but in developing electro-magnetic theory Maxwell found it on analysis, to be inadequate (something missing) in the case of momentary changing electric fields. He, therefore, introduced the highly important concept of displacement current which originated on the discharge of a condenser placed in the conduction current circuit on closing it. If J_d be displacement current density, then Eq. (9) above becomes

$$\text{Curl } B = \frac{4\pi}{c}(J + J_d) \qquad ... (10)$$

where $$J_d = \frac{1}{4\pi}\frac{\partial E}{\partial t} \qquad \textit{(see next para)}$$

$$\therefore \qquad \text{Curl } B = \frac{4\pi J}{c} + \frac{1}{c}\frac{\partial E}{\partial t} \qquad ... (11)$$

In vacuum the conduction current density J is zero, therefore Eq. (11) reduces to

$$\text{Curl } B = \frac{1}{c}\frac{\partial E}{\partial t} \qquad ... (12)$$

Evaluation of Displacement Current Density

If a quantity *dq* of charge on the plates, say of a parallel plate condenser, is discharged in a small time *dt* then the displacement current

$$i = dq/dt \qquad ...(13)$$

Now $dq =$ Pot. diff. × capacity of condenser

$$= dV \times A/4\pi x$$

where dV is the Pot. diff. between the plates of the condenser, A = the area of either plate, and x = distance between them.

∴ discharge per unit area of either plate = $dV/4\pi\, x$ and displacement current per unit area or the current density

$$J_d = \frac{dV/4\pi x}{dt} = \frac{1}{4\pi}\frac{DV/x}{dt}$$

$$= \frac{1}{4\pi}\cdot\frac{dE}{dt} \qquad \text{... (14)}$$

since $dV/x = dE$, the change in the electric intensity between the plates of the condenser. Recognizing that dE may depend on position as well as time, we shall write

$$\frac{\partial E}{\partial t} \text{ for } \frac{dE}{dt}$$

∴ Eq. (14) becomes $\mathbf{J}_d = \frac{1}{4\pi}\frac{\partial \mathbf{E}}{\partial t}$ . ... (14*a*)

The displacement current is accompanied by a momentary changing magnetic field. This momentary magnetic field will produce an induced e.m.f. which in turn produces a new displacement current. Hence if an initial displacement current is produced, an electromagnetic wave will be propagated in all directions from the centre of its origin.

(*iii*) Equation (3*c*) represents *Coulomb's law* of flux of induction which has been discussed under the head, "Gauss's theorem", where we showed that

$$\text{div } \mathbf{E} = 4\pi\rho$$

where ρ is the charge per unit volume and E, the electric field. In vacuum, $\rho = 0$,

$$\therefore \qquad \text{div } \mathbf{E} = 0 \qquad \text{... (15)}$$

or symbolically

$$\nabla . \mathbf{E} = 0$$

(*iv*) Equation (*3d*) represents that magnetic field *B* is due to currents alone. In the same way as above

$$\text{div } \mathbf{B} = 4\pi\, \delta,$$

where δ is the volume density of magnetic pole and **B** is the magnetic field.

If no isolated poles exist as in vacuum, *i.e.*, δ = 0, the net flow of magnetic induction from any enclosed surface is zero,

hence $$\text{div } \mathbf{B} = 0 \qquad \text{... (16)}$$

i.e., $$\nabla . \mathbf{B} = 0$$

Thus Maxwell's equations of electromagnetic field are derived.

Classical wave equation for e.m. wave in vacuum. Find a partial differential equation for **E** by eliminating **B** from equation (3*b*). To do so, we first differentiate equation (3*b*), *viz.*,

$$\nabla \times \mathbf{B} = \frac{1}{c} . \frac{\partial \mathbf{E}}{\partial t}$$

with respect to *t*, whence

$$\nabla \times = \frac{\partial \mathbf{B}}{\partial t} = \frac{1}{c} . \frac{\partial^2 \mathbf{E}}{\partial t^2} \qquad \text{... }(i)$$

then substitute from Eq. (3*a*) the value of

$$\frac{\partial \mathbf{B}}{\partial t} = -c\, \nabla \times \mathbf{E}$$

in Eq (*i*) above whence we have

$$-\nabla \times \ (c\nabla \times \mathbf{E}) = \frac{1}{c} . \frac{\partial^2 \mathbf{E}}{\partial t^2}$$

$$\text{or } \nabla \times \ (c\nabla \times \mathbf{E}) = \frac{1}{c} . \frac{\partial^2 \mathbf{E}}{\partial t^2} \qquad \text{...}(ii)$$

Making use of the vector identity given below :

For any vector **C**

curl (curl **C**) = Grad (div **C**) – del² **C**.

Setting E for C in the above identity, we obtain from Eq. (*ii*) above

$$\nabla\ (\nabla.E) - \nabla^2\mathbf{E} = -\frac{1}{c^2}\cdot\frac{\partial^2\mathbf{E}}{\partial t^2} \qquad ...(iii)$$

but by Maxwell's Eq. (3 *c*), $\nabla.\mathbf{E} = 0$.

Substituting zero for $\nabla.\mathbf{E}$ in Eq. (*iii*), we obtain

$$\nabla^2\mathbf{E} = \frac{1}{c^2}\cdot\frac{\partial^2\mathbf{E}}{\partial t^2} \qquad ...(17)$$

or

$$\frac{\partial^2 E}{\partial x^2} = \frac{1}{c^2}\cdot\frac{\partial^2\mathbf{E}}{\partial t^2}$$

which is the classical wave equation.

The vector **E** consists of three rectangular components E_x, Ey E_z, giving three separate partial differential equations as below :

$$\frac{\partial^2 E_a}{\partial x^2} = \frac{1}{c^2}\frac{\partial^2 E_a}{\partial x^2}, \frac{\partial^2 E_a}{\partial y^2} = \frac{1}{c^2}\frac{\partial^2 E_a}{\partial t^2}$$

and

$$\frac{\partial^2\mathbf{E}_z}{\partial z^2} = \frac{1}{c^2}\frac{\partial^2\mathbf{E}_z}{\partial t^2} \qquad ...(17a)$$

Thus E_x, E_y, and E_z each satisfies the classical wave equation for non-dispersive waves represented by

$$\nabla^2\phi = \frac{1}{v^2}\frac{\partial^2\psi}{\partial t^2},$$

where v is the velocity of the waves and ϕ their amplitude. Proceeding in the same way as above one can eliminate E from

Maxwell's equation (3*a*) by first differentiating it with respect to *t* and then substituting in it the value of $\frac{\partial \mathbf{E}}{\partial t}$ from Eq. (3*b*), we can obtain the classical wave equation for the three rectangular components of **B**. The equations show that both the intensities **E** and **B** or **H** as customarily used in radiation problems, whose variations constitute the e.m waves, travel with the same speed *c* in vacuum = 3×10^{-8} ms^{-1} a value which is the speed of light. This striking coincidence led Maxwell to predict the e.m. nature of light waves which are due to die periodic variations of electric and magnetic field intensities *E* and *B* or *H* respectively, occurring mutually at right angles to each other and also to the direction of propagation of e.m. waves, so that for an electromagnetic wave travelling in the direction *x* the intensities of the two vectors E and H are given by

$$\frac{\partial^2 \mathbf{E}_y}{\partial x^2} = \frac{1}{c^2}\frac{\partial^2 \mathbf{E}_y}{\partial t^2}$$

and $$\frac{\partial^2 \mathbf{H}_z}{\partial x^2} = \frac{1}{c^2}\frac{\partial^2 \mathbf{H}_z}{\partial t^2} \qquad ...(18)$$

respectively.

The field $\mathbf{E}_v$ and $\mathbf{H}_z$, vary sinusoidally with respect to time *t* and position *x* in phase with each other as given by the equations,

$$\mathbf{E}_y = \mathbf{E}_z \cos 2\pi f \left(t - \frac{x}{e}\right)$$

and $$\mathbf{H}_z = \mathbf{H}_0 \cos 2\pi f \left(t - \frac{x}{e}\right) \qquad ...(18a)$$

where $f = \frac{1}{2\pi\sqrt{LC}}$ is the frequency of the wave and *L* and *C* are respectively the inductance and capacitance of the circuit.

(*a*). *Electromagnetic waves in isotropic dielectric.* Consider an isotropic dielectric with electromagnetic properties of permittivity K and permeability μ. The Maxwell's equations of e.m. field will be modified as under :

$$\left.\begin{aligned}\nabla \times E &= \frac{\mu}{c}\cdot\frac{\partial H}{\partial t} \qquad \because B = \mu H \quad &...(a)\\ \nabla \times H &= \frac{K}{c}\cdot\frac{\partial E}{\partial t} &...(b)\\ \nabla . E &= 0 &...(c)\\ \nabla . H &= 0 &...(d)\end{aligned}\right\} \quad ...(19)$$

Proceed to derive the classical wave equation in the same manner, whence we have the partial differential equation for electric intensity E as

$$\nabla^2 E = \frac{K\mu}{c^2}\cdot\frac{\partial^2 E}{\partial t^2} \qquad ...(i)$$

or

$$\frac{\partial^2 E}{\partial x^2} = \frac{K\mu}{c^2}\cdot\frac{\partial^2 E}{\partial t^2}$$

Similarly, the partial differential equation for magnetic intensity H is

$$\nabla^2 H = \frac{K\mu}{c^2}\cdot\frac{\partial^2 H}{\partial t^2} \qquad ...(ii)$$

or

$$\frac{\partial^2 H}{\partial x^2} = \frac{K\mu}{c^2}\cdot\frac{\partial^2 H}{\partial t^2}$$

The equations (*i*) and (*ii*) have the form of classical wave equation expressed by

$$\nabla^2 \psi = \frac{1}{v^2}\cdot\frac{\partial^2 \psi}{\partial t^2} \qquad ...(iii)$$

where v is the velocity of propagation of the wave with ψ as its magnitude.

Hence, we find that the variations of E and H travel in a homogeneous and isotropic medium (K and μ uniform) with a velocity,

$$v = c/\sqrt{K\mu} \quad \text{...(20)}$$

For vacuum $K = \mu = 1,$

$$\therefore \quad v = c$$

the ratio of e.m u. of charge and e.s.u. of charge.

Experimentally v is found to be 3×10^{-8} ms^{-1} which is the velocity of light in vacuum.

Light waves are due to electric vector. The question arises whether it is the electric vector or the magnetic vector that constitutes the light vibrations. The answer to this question is provided when we consider the reflection of light from the surface of a transparent medium like glass on the electromagnetic theory. Several experiments beside Wiener's experiment which showed the existence of stationary light waves on a thin film of silver chloride in collodium when performed using plane polarized e.m. waves showed that the effects produced by the electric intensity vector only were similar to those produced by plane polarized light waves, thus confirming that these are the periodic changes in the electric intensity vector which constitute light waves.

In an unpolarised beam the electric and magnetic intensity vectors are rapidly rotating around the direction of propagation of the beam while in a plane polarised beam they are fixed.

Relativist invariance of Maxwell's equations. The equations are invariant in respect of Lorentz-Einstien relativistic transformation of coordinates. They have the same form for a system at rest as for a system in uniform motion in a straight line relative to it, *i.e.*, if the equations are expressed in terms of the coordinates x, y, z, and time t of the system at rest and the coordinates x', y', z', and time t' of the system in uniform motion in a straight line relative to it as modified are substituted for them the form of equations is unaltered, since the equations sum up all optical and electrical phenomena. All physical phenomena are supposed to be essentially electrical and hence they are unaffected by motion.

Electromagnetic waves and light waves compared. The theory, therefore, traces the ultimate nature of light vibrations to the atoms which radiate out energy continuously. It thus retains the essential characteristic of a wave theory and hence the lumini-ferous ether which is still the seat of variations. The mechanical or elastic solid ether is, therefore, eliminated and replaced by electromagnetic ether possessing the *electrical properties of a dielectric and magnetic permeability.* The success of this field theory lay in the fact that it did not require ether of all contradictory properties but only having such properties as are involved in Maxwell's equations. The theory explains all The optical phenomena such as reflection, refraction, interference, diffraction and polarisation and has brought together by an underlying fundamental concept the two highest unrelated branches of Physics, 'Optics' and 'Electromagnetism'. Maxwell in the original version of the theory of e.m. waves retained Seminiferous ether as pervading all space (Relativity came about 50 years afterwards) and calculated the velocity of e.m. waves as :

$$v = c/\sqrt{K\mu} \text{ cm/sec}$$

where c is the number of e.s.u. of electrical charges in 1 e.m u. of charge, K is the permittivity of ether and μ its permeability. In vacuum, both K *and* μ are 1 each ; in CGS system, therefore, $v = c = \times 10^{10}$ cm/sec. The theory also explains dispersion, that is, variation of refractive index with wavelength. The refractive index of a substance on this theory is given by

$$n = \frac{c}{v} = \frac{\text{velocity of light in vacuum}}{\text{velocity of light in the substance}}$$

$$= c \Big/ \frac{c}{\sqrt{K\mu}} = \sqrt{K\mu} \qquad ...(21)$$

For a transparent substance, $\mu = 1$ so that $n = \sqrt{K}$. Both n and K can be easily determined in the laboratory, n by optical methods and K by electrical methods.

It has been found that for many substances the agreement $n = \sqrt{/K}$, is remarkable (see table below) but for others there is a big divergence. This explained as being due to the fact that n is determined for light oscillations of high frequency while K is found under static conditions. This discrepancy will disappear, if K is also found for highly oscillating currents.

Table showing agreement $nc = \sqrt{K}$

Substance	n	$\sqrt{K}$
Hydrogen	1.000138	100032
Air	1.000294	1.000295
CO_2	1.000449	1.000473
CO	1.000346	1,000345

The only difference between the light waves and electromagnetic waves is that the former are of extremely short wavelengths.

In 1887 Hertz actually produced and detected e.m. waves. His initial experiments consisted in showing that if a thick copper wire bent into a circle with knobs at the ends separated by a small air gap is placed in the neighbourhood of a circuit through which an induction coil is sending a spark discharge, then a spark also passes in the air gap. This would occur for certain definite dimensions of the detector implying resonance effect. By improving his apparatus he was successful in showing that e.m. waves could be reflected, refracted, diffracted and polarised and had the same velocity as the velocity of light, thus experimentally establishing the truth of Maxwell's predictions. In brief, the waves behaved exactly as light waves and as Larmour remarked, *"The discoveries of Hertz left no further room for doubt that the physical scheme of Maxwell constituted the formulation of the underlying unity in physical dynamics".*

Signalling by electromagnetic waves. One sketch of apparatus for producing *e.m.* carrier wave, in which by means of an induction coil a series of sparks take place between the knobs of *S* and at each discharge Toscillations occur in the circuit consisting of condenser C and one winding of transformer *T* the other winding of which is in series with the area or antenna *A* in which it produces an oscillating e.m.f. When the periods of the two circuits agree, oscillations of large amplitude result and the antenna becomes the seat of radiations of the type upper half propagates, since lower point of the antenna is earthed. The signal is made to ride on this carrier wave. Since it is of large frequency, it is either to be frequency modulated (F.M.) or amplitude modulated (A.M.) so that the signal may be deciphered. At the receiving station the modulated e.m. carrier wave meets a similar antenna and electric surgings similar to those at the transmitting station are set up in it. Radio and Television are its everyday applications.

Electromagnetic plane waves are transverse. Suppose electromagnetic plane waves, which consist of space and time dependent electric and magnetic fields $E\ (x, y, z, t)$ and $B\ (x, y, z, t)$ are travelling in the direction x. In such waves none of the components or E or B depends on either of the transverse coordinates y and z.

Thus we have

$$E = \hat{x}E_x\ (x, t) + \hat{y}E_y\ (x, t) + \hat{z}E_z\ (x, t) \qquad ...(22)$$

$$B = x\hat{B}_x\ (x, t) + y\hat{B}_y\ (x, t) + z\hat{B}_z\ (x, t) \qquad ...(23)$$

To prove that the waves are transverse we apply Maxwell's equations to Eqns. (22) and (23) above.

First we use Gauss's theorem, $\nabla . =0$. Also, since there are no y and z dependences of any one of the components, their partial derivatives with respect to y and z are zero.

Thus, we have

$$\nabla. E = \frac{\partial E_x(x,t)}{\partial x} = 0 \qquad ...(24)$$

since $$\nabla. E = \hat{x}\frac{\partial E_x}{\partial x} + \hat{y}\frac{\partial E_y}{\partial y} + \frac{\partial E_z}{\partial z} \text{ -}= 0$$

by Eq. (14 *a)* of Appendix V.

Therefore E_X is independent of x. To show that E_X is also independent of t, we consider Maxwell's "displacement current'" equation,

$$\nabla \times B = \frac{1}{c}\cdot\frac{\partial E}{\partial t} \qquad ...(25)$$

Take the x component of Eq. (25) whose left-side involves $\frac{\partial B_z}{\partial y}$ and $\frac{\partial B_y}{\partial y}$, since $\nabla \times B = +\hat{x}\left(\frac{\partial B_z}{\partial y} - \frac{\partial B_y}{\partial z}\right)$ $+\hat{y}\left(\frac{\partial B_x}{\partial z} - \frac{\partial B_z}{\partial x}\right) + \hat{z}\left(\frac{\partial B_y}{\partial x} - \frac{\partial B_x}{\partial y}\right)$. Both of them are zero, by the definition of the plane wave, therefore their equivalent $\frac{1}{c}\ \frac{\partial E_x}{\partial t}$ on the right side and hence $\frac{\partial E_x}{\partial t}$ is on the right side and hence zero *i.e.* E_a is constant. For simplicity we take this constant to be zero without losing generality. Similarly, using equation $\nabla B_X = 0$ we can show that $B_{x.}$ is independent of x And it also has no time dependence is seen by considering Faraday's law

$$\nabla \times E = \frac{1}{c}\cdot\frac{\partial B}{\partial t} \qquad ...(26)$$

which gives $\delta B_X/\delta t$ to be zero. Therefore B_x may be taken as zero. Thus we conclude that *electromagnetic plane waves are transverse waves.* That is, electric and magnetic fields are perpendicular to the direction of propagation $\hat{x}$.

Polarization of plane electromagnetic waves. Let us first define what is a state of polarization. We have already learned that electric and magnetic fields in plane electromagnetic waves are transverse to $\hat{x}$, the direction of propagation of the waves. There are two transverse directions $\hat{y}$ and $\hat{z}$ and the fields with one orientation with respect to $\hat{y}$ and $\hat{z}$ are independent of those with an orientation differing by $\pi/2$. One may therefore have various amplitudes of thc fields in each of the two transverse directions and the various possible relative phases. A *specific* relation of the amplitudes and phases of the two independent transverse fields is called a *state of a polarization.*

Let us apply this criterion of polarization to electromagnetic plane transverse waves having **E** (x, t) as its transverse electric field. The other quantity which concerns us is the transverse magnetic field **B** (x, t) which is known being dependent on known $E(x, t)$. For example, we can resolve a general **E** (x, t) in the superposition of travelling waves, travelling in both $+x$ and $-x$ directions. Let E^+ denote the part of E contributed by waves travelling in the $+x$ direction and E^- the part contributed by waves in the $-x$ direction so that

$$E(x, t) = E^+(x. t) + \mathbf{E}^-(x, t) \qquad \dots (27)$$

It can be shown that in travelling waves the magnetic field B^+ corresponding to E^+ is equal to $\hat{x} \times E^+$ vide vectorial rule, $\hat{z} = \hat{x} \times \hat{y}$ and the value of $\mathbf{B}''$ corresponding to E^- is equal to $-\hat{x} \times E^-$. Thus the magnetic field corresponding to the above equation is

$$B(x, t) = \hat{x} \times [E^+(x, t) - E^-(x, t)] \qquad \dots (28)$$

The condition of the state of polarization stated in the beginning is, therefore, satisfied in the case of electromagnetic plane transverse waves.

Linear polarization. In the case of plane electromagnetic waves if the displacement is an oscillation back and forth along a fixed line transverse to $\hat{x}$, the direction of propagation, the waves are said to be *linearly polarized.* There are two independent transverse direction $\hat{y}$ and $\hat{z}$. Let us take a fixed value of x. Then the oscillations corresponding to a linearly polarized plane wave can be expressed by the equations.

$$\psi(x, t) = \hat{y}\, a \cos \omega t \qquad \text{... (29)}$$

and $$\psi(x, t) = \hat{z}\, b \cos \omega t \qquad \text{... (30)}$$

where phase constant is zero.

More generally, the linearly polarized oscillation may not be along $\hat{y}$ or $\hat{x}$ axis. Such a vibration can be expressed by the superpositions of the vibrations denoted by Eqns. (29, 30) with the same *phase constant.*

$$\psi(x, t) = \hat{y}\, a \cos \omega t + \hat{z}\, b \cos \omega t \qquad \text{... (31)}$$

or $$\psi(x, t) = (\hat{y}\, a + \hat{z}\, b) \cos \omega t \qquad \text{... (31 } a\text{)}$$

the vector $\hat{y} a - \hat{z}\, b$ has magnitude and direction that are independent of time. Therefore $\psi(x, t)$ as given in Eq. (31 a) consists of an oscillation along a fixed line. The amplitude A of the oscillation is given by

$$A = \sqrt{a^2 + b^2} \qquad \text{... (32)}$$

The direction of the linear polarization makes an angle θ with y-axis such that

$$\tan\theta = \frac{b}{a}$$

or $$\theta = \tan^{-1}\frac{b}{a} \qquad \text{... (33)}$$

The displacement $\psi(x, t)$ for a linearly polarized wave (at fixed x).

where $\hat{x}$, the direction of propagation of the wave is perpendicular to the plane of the paper through O.

Circular polarization. If the displacement in transverse waves is a motion in a circle, the waves are said to be *circularly polarized.*

We first consider the state of the waves at a fixed value of x, a point on the direction of propagation of the wave. If the thumb of your right-hand points along $+\hat{x}$ direction when the fingers curl in the direction of rotation, then the oscillation is said to be circularly polarized along $+\hat{x}$. Similarly, we use right-hand rule to define circular polarization along $-\hat{x}$. An oscillation circularly polarized along $+\hat{x}$ can be expressed as a superposition of linearly polarized oscillations along $\hat{y}$ and $\hat{x}$ axes with the same amplitude as the oscillation along $\hat{y}$. We thus see that for circular polarization along $+\hat{x}$, the $\hat{y}$ oscillation *leads* the $\hat{z}$ oscillation by $\pi/2$.

$$\psi(x, t) = \hat{\mathbf{y}}\, a\cos\omega t + \hat{\mathbf{z}}\, a\cos\left(\omega t - \frac{\pi}{2}\right) \qquad \text{...(34)}$$

$$= \hat{\mathbf{y}}\, a\cos\omega\, t + \hat{\mathbf{z}}\, a\sin\omega t \qquad \text{... (35)}$$

Similarly, for circular polarization along $-\hat{x}$, the $\hat{y}$ oscillation *lags* the $\hat{z}$ oscillation by $\pi/2$.

$$\psi(x, t) = \hat{\mathbf{y}}\, a \cos \omega t + \hat{\mathbf{z}}\, a \cos\left(\omega t + \frac{\pi}{2}\right) \quad ...(36)$$

$$= \hat{\mathbf{y}}\, a \cos \omega t — \hat{\mathbf{z}}\, a \sin \omega t \quad ...(37)$$

The displacement $\psi(x, t)$ for circularly polarized oscillation is $\hat{x}$, the direction of propagation of the wave is perpendicular to the plane of the paper through O.

General transverse polarization—elliptical polarization. At a fixed value of x the direction of propagation of a plane wave a general transversely polarized oscillation has the form

$$\psi\ (x,\ t) = \hat{\mathbf{y}}\ \text{a} \cos\ (\omega t + \phi_1) + \hat{\mathbf{z}}\ b \cos\ (\omega t + \phi_2) \quad ...(38)$$

If ϕ_2 equals either ϕ_1 or $\phi_2 + \pi$, we have linear polarization along $+x$. If ϕ_2 is $\phi_1 + \pi/2$ and b equals a, we have circular polarization along $-\hat{x}$. For the general case where a and b are unequal and ϕ_1 and ϕ_2 are arbitrary, the displacement $\psi(x, t)$ describes an elliptical path. Let ψy and ψz be called the y and z displacements, then y is $a \cos (\omega t + \phi_1)$ and z is $b \cos (\omega t + \phi_2)$. For simplicity, let the two rectangular vibrations be represented by

$$y = \cos\ (\omega t + \phi) \quad ...(i)$$

and

$$z = b \cos \omega t \quad ...(ii)$$

To find the equation of the resultant motion as a consequence of their superposition let us eliminate ω from (*i*) and (*ii*).

From (*i*) by expanding cosine term

$$\frac{y}{a} = \cos \omega t \cos \phi — \sin \omega t \sin \phi \quad ...(iii)$$

Putting the values of $\cos \omega t$ and $\sin \omega t$ from (*ii*) in (*iii*) above" we get

$$\frac{y}{a} = \frac{z}{b} \cos \phi - \sqrt{1 - \left(\frac{z}{b}\right)^2} \sin \phi$$

or $$\frac{y}{a} - \frac{z}{b} \cos \phi = -\sqrt{1 - \left(\frac{z}{b}\right)^2} \sin \phi$$

squaring and simplifying

$$\frac{y^2}{a^2} + \frac{z^2}{b^2} \cos^2 \phi - \frac{2yz}{ab} \cos \phi = \sin^2 \phi - \frac{z^2}{b^2} \sin^2 \phi$$

or $$\frac{y^2}{a^2} + \frac{z^2}{b^2} (\cos^2 \phi + \sin^2 \phi) - \frac{2yz}{ab} \cos \phi = \sin^2 \phi$$

or $$\frac{y^2}{a^2} + \frac{z^2}{b^2} - \frac{2yz}{ab} \cos \phi = \sin^2 \phi \qquad \text{... (39)}$$

which is the general equation, of an ellipse. The changes as we vary relative phase ϕ in Eq. (39).

Later Zeeman, Stark and Kerr effects were discovered which lent a further support to the theory.

Zeeman effect. Prof. P. Zeeman, a Dutch spectro-scopist, discovered in 1896 that if a source of light producing line spectrum, say, for example a sodium flame is placed between the poles of a strong electromagnet and the light emitted is examined through a spectrometer of high resolving power each spectral line is found to split up into two or more components. If the light is viewed in a direction at right angles to the field the original line is accompanied by two other lines, one on each side of it. *The central component becomes plane-polarised with displacement parallel to the field and the two outer components being plane polarised at right angles to the central one.* This is called the Transverse Zeeman effect.

If the light is viewed along the direction of the field by

boring a hole in one of the pole pieces, *the central line disappears and the two outer components are only seen, both of them being circularly polarised in opposite directions.* This is known as the Longitudinal Zeeman effect.

Explanation on electronic theory of matter. Soon after the discovery, Lorentz gave a simple explanation of the phenomenon. He based his theory on the assumptions that (*i*) the motion of the extra nuclear electrons of an atom is responsible for the emission of light by a glowing gas, (*ii*) the number of revolutions per second will determine the number of waves per second and (*iii*) this motion is modified by an applied magnetic field. Suppose we are looking down on an electron (– ve charge) revolving in a counterclockwise direction, in a horizontal plane. If a magnetic field acting vertically downward is set up, the orbit of the electron will expand (Fleming's left-hand rule), the orbital velocity of the electron, however, remains constant number of revolutions will be made per second. By a similar reasoning an electron revolving in a clockwise direction, will have its orbit diminished and will, therefore make a greater number of revolutions/sec. The orbit of an (*a*) Anti-clockwise (*b*) Clockwise electron revolving in a vertical plane, *i.e.*, in a plane paralled to the magnetic field will remain unaffected.

Polarisation effects. To explain the polarisation effects on this theory, let us consider a single electron moving around an atom. All possible motions of this electron at any instant can be resolved into a linear motion, say, along X-axis and two opposite circular rotations executed in the same period at right angles to it, *i.e.*, in the plane *YOZ*.

Transverse effect. If now a magnetic field is applied in the direction of *X* axis the *x* vibration will remain unaffected with the original frequency ν and the two opposite circular motions in the plane *YOZ*, will have altered frequencies, $\nu + \delta\nu$ and $\nu - \delta\nu$, so that on looking at *right angles* to the field which is parallel to *OX*, the linear motion along the field will send out vibrations parallel to the field of original period, say T_0, so the spectroscope will reveal a line in the undisturbed position which is plane-polarised, the vibrations being parallel to the field. The two opposite circular motions at right angles to the field being seen edge-wise will send out two linear vibrations in that direction, since it is in their own plane, one with increased and the other with diminished period as a result of the application of the field. Thus the spectroscope will show *two additional* lines, one on each side of the undisturbed line with frequencies $\nu + \delta\nu$ and $\nu - \delta\nu$. These two additional lines are also plane-polarised, the vibrations being perpendicular to the field. Hence the undisturbed line and the doublet are polarised in perpendicular planes, and they have been found to be so by Nicol's prism.

Longitudinal effect. Next imagine the light to be viewed in the direction of the field. The linear vibration along the field cannot be propagated in that direction as the displacements of the Faraday's tubes of force emanating from the electron are taking place in the perpendicular direction. The two opposite circular motions referred to above being at right angles to the field will result in the emission of two lines of opposite circularly polarised light displaced equal distances from the position of no-field line, which is absent. These lines will be of equal intensity since there are as many components of motion in one as in the other. They can be rendered plane-polarised, by a quarter-wave plate and tested by Nicol's prism.

Mathematical theory of Zeeman effect. Suppose an electron of mass *m* and charge *e* rotates in a circular orbit of radius *r* about its nucleus of charge Ze. If v is the linear velocity and ω the angular velocity of the electron in the absence of the field, then the centripetal force

$$F = \frac{1}{4\pi \epsilon_0} \times \frac{Ze^2}{r^2} = \frac{mv^2}{r} = mr\omega^2.$$

When a magnetic flux density of strength B is applied we have an additional radial force Bev as explained above either towards or away from the centre. If ω and ω_2 be the angular velocities in the two cases, we have

$$mr\omega_1^2 = mr\omega^2 + Bev_1$$
$$= mr\omega^2 + Ber\omega_1$$

and $$mr\omega_2^2 = mr\omega^2 - Ber\omega_2$$

Simplifying, we get

$$\omega_1^2 = \omega^2 + \frac{Be}{m}\,\omega_1,$$

and $$\omega_1^2 = \omega^2 - \frac{Be}{m}\,\omega_2$$

Subtracting, we have

$$\omega_1^2 - \omega_2^2 = \frac{Be}{m}(\omega_1 + \omega_2)$$

or $$\omega_1 - \omega_2 = \frac{Be}{m}.$$

On classical theory we assume that the frequencies of the light radiated are equal to the orbital frequencies, ν, ν_1, and ν_2 of the electron, therefore, we have

$$\omega_1 = 2\pi\nu_1 \text{ and } \omega_2 = 2\pi\nu_2.$$

Hence, we can write

$$2\pi(\nu_1 - \nu_2) = \frac{Be}{m},$$

or $$\nu_1 - \nu_2 = \frac{1}{2\pi} \times \frac{Be}{m},$$

or $$2\delta\nu = \frac{1}{2\pi} \times \frac{Be}{m},$$

where $\delta\nu$ is the change in the frequency of vibration from the no-field line.

$$\therefore \qquad \delta\nu = \frac{1}{4\pi}\frac{Be}{m} \qquad \text{... (40)}$$

where e is the charge on the electron in coulombs ; B, the field strength (magnetic induction) in teslas; and m the mass of the electron in kg.

Also since $v = \dfrac{c}{\lambda}$, where c is the velocity of light, we have by differentiating,

$$\delta\nu = -\frac{c}{\lambda^2}\,\delta\lambda,$$

or the Zeeman change in wavelength (Zeeman shift) is *numerically* given by

$$\delta\lambda = \delta v\,\frac{\lambda^2}{c}$$

$$= \frac{B}{4\pi c}\times\frac{e}{m}\lambda \qquad \text{... (41)}$$

If the value of $\delta\lambda$ is determined for a given magnetic field, we can calculate e/m. Runge and Paschen found that or a certain series of mercury lines e/m came out to be 1.65×10^{11} C kg^{-1}.

This value is in close agreement with that obtained by purely electrical methods thus confirming the truth of the picture regarding the electromagnetic origin of spectra.

The Zeeman effect gave a strong confirmation of the electromagnetic theory of light for it followed from that theory that whenever an electric charge oscillated, it would send out an electric wave (light). A magnetic field should influence the oscillations in such a way that in a direction perpendicular to the field a spectral line should be a triplet and along the field it should be a doublet; that the line should now be polarised, whereas the original line was unpolarised. Furthermore, soon after the discovery of the electron and the determination of *e/m* the displacements of the components of the spectral line were quantitatively verified.

Example. *Calculate the Zeeman charge in wavelength (or Zeeman shift) in the case of sodium light of wavelength 6000 Å in a field of 10000 gausses, given e/m = 1.759 × 10¹¹ coulomb/kg.*

$$\text{Magnetic field} = 10000 \text{ gausses} = 1 \text{ weber/m}^2,$$

$$\lambda = 6 \times 10^{-7} \text{ m} \quad [\because 10^4 \text{ gauss} = 1 \, wb \text{ m}^{-2}]$$

Now Zeeman shift,

$$\delta\lambda = \frac{B}{4\pi c} \cdot \frac{e}{m} \lambda^2.$$

Substituting the values, we have

$$\delta\lambda = \frac{1}{4\pi \times 3 \times 10^8} \times 1.759 \times 10^{11} \times (6 \times 10^{-7})^2$$

or
$$\delta\lambda = \left(\frac{1.759 \times 36}{12 \times 3.142}\right) 10^{-11}$$

$$= 0.168 \times 10^{-10} \text{ m}$$

$$= 0.168 \text{ Å}.$$

Faraday effect. Faraday discovered in 1845 that a transparent isotropic substance when placed in a strong magnetic field, acquires the property of rotating the plane of polarisation of a beam of light which is sent through it in the direction of the field. This effect is called Faraday effect after the name of its discoverer. The behaviour is similar to *optically active* substances like quartz and sachharine solutions in so far that a rotation of the plane of polarisation occurs but there is one important difference. In this case the direction of rotation depends upon whether the light passes the medium in the direction of the lines of force or oppositely. Hence if the light is reflected back through a medium in a strong magnetic field the rotation of the plane of polarisation is double while the rotation due to an optically active substance is neutralised if the ray is reflected back on its path.

An arrangement for studying Faraday effect, the substance contained in a tube T is placed on the axis of a solenoid, and a beam of light is sent through it from left to right When the nicols N_1 and N_2 are crossed in the absence of the field, no light passes through but the light reappears when the field is on. The angle through which the analysing nicol N_2 is rotated to extinguish the light gives a measure of the rotation of the plane of polarisation. It is found that the amount of rotation is proportional to the field strength and to the distance the light travels through the medium.

For a given temperature and wavelength the rotation θ_r may be expressed in minutes of an arc by the relation

$$\theta_r = VlH \qquad \dots (9)$$

where l is the light path in cm in the medium H, the field strength in gausses and V a constant depending upon the nature of the medium known as Verdet's constant. It is defined as the rotation per unit path per unit field strength. V is *positive* if the direction of rotation is the same as that of the current which is producing the magnetic field and *negative* if in the opposite direction. The effect has been measured in a large number of substances including liquids and gases. For explanation of Faraday effect and several other magneto-and electro-optic effects such as Kerr effect. Cotton-Mouton effect, Voigt effect, the student is advised to consult some advanced book on Optics or Modern Physics.

Stark effect. In 1913 J. Stark of Germany observed that when a hydrogen discharge tube is subjected to a powerful electric field and the spectrum is examined by a high resolving power spectroscope a single line is split up into several components symmetrically situated. This effect is analogous to

Zeeman's discovery of the splitting up of spectral lines in a magnetic field and is known as Stark effect. When the spectrum is viewed perpendicularly to the electric field some of the components of each line pattern are observed to be plane-polarised in the plane of the field and some others in a perpendicular direction. This is the *transverse* Stark effect. When viewed parallel to the field only the latter type of components appear but as ordinary unpolarised light. This is the *longitudinal* stark effect. A satisfactory explanation of this effect is based on Quantum theory and may be considered as beyond the scope of this book.

Dispersion. *Normal dispersion.* It has been observed that in general the refractive index of a medium decreases regularly with increasing wavelengths and the rate of decrease becomes smaller at longer wavelengths ; thus the curves for substances with larger values of refractive index are steeper. This is quite true for wavelengths far removed from the absorption bands of the medium. This is termed as normal dispersion. The curves for normal dispersion of many transparent media in the visible region can be represented quite accurately by Cauchy's equation.

$$\mu = A + \frac{B}{\lambda^2} + \frac{C}{\lambda^4} \qquad ...(1)$$

where A, B and C are constants which are characteristic of any substance. Generally this equation with two constants is used, that is,

$$\mu = A + \frac{B}{\lambda^2}.$$

From which dispersion,

$$\frac{d\mu}{d\lambda} = -\frac{2B}{\lambda^3} \qquad ...(2)$$

The minus sign indicates the negative slope of the dispersion curve.

Anomalous dispersion. Cauchy's equation predicts a very gradual decrease of refractive index for large values of wavelength. But actually it has been found that the value of refractive index decreases much more rapidly as it approaches the ıegion of absorption band, the position of which is characteristic of the material. Beyond the absorption band refractive index is found to be very high and it decreases rapidly at first and then slowly as we proceed towards the longer wavelength side. Cauchy's equation holds good again in this region.

This discontinuity in the dispersion curve as it crosses the absorption band gives rise to what is known as anomalous dispersion. In this small wavelength region refractive index increases rapidly with increase in wavelength. However in this region refractive index cannot be measured as light is not transmitted through the medium. For example Å in the case of sodium vapour in the visible region anomalous dispersion has been observed at wavelengths 5890 Å and 5896 Å. Similarly iodine vapour also shows anomalous dispersion in the visible region.

The curves *AB* and *CD* show normal dispersion whereas in the absorption region, where μ increases rapidly with Å, this represents anomalous dispersion. Cauchy's equation could very well explain normal dispersion but it failed to explain anomalous dispersion, whereas Sellmeir's formula could explain both normal as well as anomalous dispersion. His formula for μ of a medium is given by

$$\mu^2 = 1 + \sum_p \frac{A_p \lambda^2}{\lambda^2 - \lambda_p^2} \qquad ...(3)$$

where A_p is proportional to the number of particles/unit volume vibrating with a natural frequency corresponding to wavelength A, and A is the wavelength of incident radiation.

In case the medium has two absorption wavelengths corresponding to λ_1 and λ_2, the above relation reduces to

$$\mu^2 = 1 + \frac{A_1 \lambda^2}{\lambda^2 - \lambda_1^2} + \frac{A_2 \lambda^2}{\lambda^2 - \lambda_2^2} \quad \text{... (4)}$$

It is obvious that as λ approaches λ_1 or λ_2 from the longer wavelength side μ tends to $+\infty$ at resonance whereas if λ approaches λ_1 or λ_2 from shorter wavelength side μ tends to $-\infty$ at resonance.

When white light is passed through sodium vapours, there is a strong selective absorption in the neighbourhood of D_l and D_2 lines. For showing this anomalous dispersion of sodium vapour an experiment was devised by Prof. R.W. Wood.

Wood's Experiment

In its simple form the apparatus consists of a cylindrical steel tube *T* fitted with water-cooled glass-windows W_1 and W_2 at the ends. S is a source of white light, L_1 is a collimating lens and a spectrograph is placed on the other side of the tube.

Pieces of metallic sodium are placed along the bottom of the whole length of the tube which is partially evacuated and heated slowly by burners from below. Sodium vaporizes and its density decreases from bottom to upwards. Thus a density gradient is set up from the bottom to the top of the tube, which is equivalent to a prism of sodium vapours. When white light is passed through the tube, a spectrum is obtained on the photographic plate in the spectrograph. It is observed that the spectrum is continuous except at position of D_1 and D_2 lines indicating the presence of absorption bands at these positions.

6

Polarisation of Waves

Rotatory Polarisation

It is found that if plane polarised light be made to pass through certain substances the plane of polarisation of the emergent light is not the same as that of the incident light but has been rotated through a certain angle. This phenomenon is known as rotatory polarisation and this property of crystals and other substances is called optical activity and the substances which rotate the plane of polarisation are said to be *optically active.*

There are two types of optically active substances—those which produce clockwise rotation (when one looks towards the light source) are known as dextro-rotatory *(right handed)* substances-and others which rotate, the plane of polarisation in the anti-clockwise direction are called laevo-rotatory *(left-handed)* substances.

If monochromatic light from a source *S,* is made to pass through two nicols *PN* and *AN* put in a *conventional* crossed position, there is no emergent light. The polarising, nicol *PN* renders the beam plane-polarised with its vibrations in the principal plane. The analysing nicol *AN* in this position is set with its principal plane perpendicular to the direction of these vibrations; so that no fight is allowed to pass through it. If a quartz crystal cut *with its refracting faces perpendicular to the optic axis* is placed between the two nicols, still crossed, and light is incident normally on its faces, it is found that light is no longer cut off by the analysing nicol. To produce extinction of light the analysing nicol is required to be rotated through some angle. Thus the emergent light is still plane-polarised but its plane of polarisation has been rotated.

If the crystal is turned around so that the beam of tight passes through it in opposite direction, no change in the direction of rotation of the plane of polarisation is observed. Experiments have farther shown that some quartz crystals *are left-handed* and some right handed. Fused quartz is, however, optically inactive showing thereby that the optical activity is due to the crystalline structure of the substance.

Many liquids and organic substances in solution have been found to be optically active. Turpentine oil, aqueous solutions of various of sugars and of tartaric acid, etc., all cause the plane-polarised tight to be rotated through an angle which is proportional to the thickness of the medium which the light has to pass through.

The amount of optical rotation depends upon the thickness and density of the crystal or concentration in case of solutions, the temperature and the wavelength of light employed.

Fresnel's Explanation of Optical Rotation

Fresnel's explanation of optical rotation is based upon the assumption that when plane-polarised light is allowed to pass through a crystal *along the optic axis* it ft decomposed into two circulatory polarised vibrations rotating in opposite directions with the same frequency. In a crystal-like calcite which is not optically active, these two circular motions travel with the same speed and starting from the point of intersection of YY' with the circle arrive at the equidistant points, P_1 and P_2, at the same instant and in the same phase. Since any given point is their path is simultaneously arrived at it is obvious that at emergence they give rise to a linear vibration in the same plane as that of the incident vibration. In optically active crystals these two circularly polarised vibrations travel with unequal speeds. As a consequence at emergence, the two circularly polarised; components have a certain phase difference between them, and combine; to form plane-polarised light whose direction of vibration will be inclined to the primitive plane of polarisation. The rotation of the plane of polarisation will thus be accounted for by the retardation of a circular component.

In a *right-handed quartz*, right-handed motion travels Caster than the *left-handed* while in a *left-handed quartz* left-handed motion travels faster than the right-handed. As a result of this the rotation of the vibration, plane is to the right or left according as the right-handed or the left-handed component is faster and is equal to half the phase difference on emergence between the two circular vibrations. These results may be arrived at very easily by means of the equations of the vibration. Since each circular motion is equivalent to two rectangular simple harmonic motions, so in a crystal which is not optically active the components of the right-handed circular motion beginning with the point of intersection of the circle with YY^r axis are

$$x_1 = a \cos \theta = a \cos \omega t$$

$$y_1 = a \sin \theta = a \sin \omega t.$$

The components of the left-handed circular motion are

$$x_2 = -a \cos \omega t, \; y_2 = a \sin \omega t.$$

Thus the two circular motions, one right-handed and the other left-handed are equivalent to four simple harmonic motions whose resultant displacements along the x-axis and y-axis are

$$x = x_1 + x_2 = a \cos \omega t - a \cos \omega = 0$$

and

$$y = y_1 + y_1 = a \sin \omega t + a \sin \omega t$$

$$= 2a \sin \omega t.$$

Hence the x components of the circular vibrations by the principle of superposition of wave motions cancel out and the resultant motion of the particle is simple harmonic being the sum of y-components of amplitude $2a$ executed along the Y-axis. This shows that the two oppositely circularly polarised vibrations rotating with equal frequency give rise to a plane-polarised vibration. The converse of this statement is equally true and has been also-verified experimentally by Fresnel with the help of his compound prism.

Now suppose the plane-polarised light passes through a dextrorotatory substance, then the right-handed circular motion being quicker, at emergence, starts from P_2 say, while the left-handed has not yet reached it but has attained a position P_1, being retarded by an angle ϕ from P_2. The equations of the transmitted component motions are

$$x_1 = a \cos \omega t \ ;$$

$$y_1 = a \sin \omega t \qquad \text{... (right-handed)}$$

$$x_2 = - a \cos (\omega t + \phi) \ ;$$

$$y_2 = a \sin (\omega t + \phi) \qquad \text{... (left-handed)}$$

The resultant displacements along the two axes are:

$$x = x_1 + x_2 = a \cos (\omega t) - a \cos (\omega t + \phi t)$$

$$= 2a \sin (\omega t + \phi/2) \times \sin (\phi/2)$$

and

$$y = y_1 + y_2 = a \sin (\omega t) + a \sin (\omega t + \phi)$$

$$= 2a \sin (\omega t + \phi/2) \times \cos (\phi/2).$$

These two rectangular vibrations being in the same phase when compounded together give a resultant linear vibration along *CD* which makes an angle δ with the *Y-axis*, such that

$$\tan \delta = x/y = \tan (\phi/2).$$

i.e.,

$$\delta = \phi/2.$$

Thus the vibration plane of the incident plane-polarised light has been rotated in the clockwise direction by an angle which is equal to half of the phase difference between the two emergent circularly polarised vibrations.

We can prove the same relation for *a left-handed* substance.

Calculation of the angle of rotation. The magnitude of ft the angular rotation may be calculated in terms of the refractive indices of the right-handed and left-banded vibrations. If V_2 and V_1 be the velocities of the right-handed and left-handed vibrations, t the thickness of the crystal along the optic axis, the time retardation between the two circular components in traversing the right-handed quartz is

$$\Delta T = T_L - T_R = \frac{t}{V_L} - \frac{t}{V_R}.$$

If V_a be the velocity of light in air the path retardation

$$= V_a \Delta T = t\left[\frac{V_a}{V_L} - \frac{V_a}{V_R}\right] = (\mu_L - \mu_R)$$

where μ_L and μ_R are the refractive indices of the quartz for propagation along the optic axis of *L* and *R* circularly polarised component of the incident plane-polarised light.

Hence the phase difference between the two circular components

$$\phi = \frac{2\pi}{\lambda} t (\mu_L - \mu_R)$$

∴ the angle of rotation of the plane of vibration by the right-handed quartz is given by

$$\delta = \frac{\phi}{2} = \frac{\pi t}{\lambda} (\mu_L - \mu_R). \quad \text{... (1)}$$

Similarly, the angle of rotation of the plane of vibration by the left-handed quartz is expressed by

$$\delta = \frac{\phi}{2} = \frac{\pi t}{\lambda} (\mu_R - \mu_L). \quad \text{... (2)}$$

Experimental verification of Fresnel's theory of rotatory polarisation. To verify his assumption experimentally Fresnel constructed a compound rectangular block made up of several right-handed and left-handed quartz prisms placed together alternately right and left. In each the optic axis was *parallel* to its respective base.

When a plane polarised beam is incident normally on the face *AB* of the block, it is decomposed into two component circular vibrations which are propagated along the optic axis with different speeds, right handed motion being faster than left-handed. On crossing *BC* the beam enters the left-handed quartz so that the right-, handed motion becomes slower and the left-handed faster. For the *R* component, therefore, the prism *L* behaves a» a *denser* medium so by the ordinary law of refraction, this component is refracted towards the boundary normal and the *L* component away from it, for with respect to it the medium is *rarer*. At every boundary of the prisms the velocities are inter-changed. So that the ray bending towards the normal at one boundary bends away from it at the other. The net' result is that the angular separation of the two circular vibrations increases at each successive refraction and finally two distinctly separate beams emerge out of the block. On analysing the emergent beams with the usual device of a quarter-wave plate and a nicol, it is established that they are circularly polarised beams, one rotating to the right and the other to the left.

Specific Rotation

To bring the rotation of all optically active substances into a comparable form the term specific rotation has been adopted.

'Specific rotation', at a given temperature t and for a given wavelength of light λ is defined as the rotation (in degrees) produced by a path of one decimeter length in a substance of unit density.

If θ is the rotation produced by *l* decimetre length of a solution of density *d/cc*, then the specific rotation S at a given temperature *t* and for a given wavelength of light λ used is given by

$$S_t\lambda = \frac{\theta}{I \times d} \qquad ...(3)$$

i.e., specific rotation =

$$\frac{\text{rotation in degrees}}{\text{length in decimetres} \times \text{concentration in gm / cc.}}$$

Molecular rotation is given by the product of the specific rotation and the molecular weight of the substance. It is defined as the rotation produced by one decimetre length of solution containing one Gram-molecule of the active substance per cc of the solution.

Sugar is probably the most common of all the optically active substances and this optical activity is generally, employed for the estimation of its strength in a solution by measuring the rotation of the plane of polarisation. The instruments designed for determining the optical rotation produced by & substance are called polar-meters or saccharimeters. Essentially they consist of two nicols placed some distance apart and capable of rotation about the incident beam as axis. The nicols are first adjusted to be in the crossed position so that the field of view is dark. Then a tube containing the solution it introduced between the nicols, when it would be found that the field of view is no longer dark. To restore the position of extinction the analysing nicol it rotated through a certain angle θ demanding upon the length *of* the tube and the nature of the substance. Knowing θ, *l* and *d* the specific rotation is then calculated from the above, formula. We describe; below a few improved forms of polarimeters.

Laurent's Polarimeter

The essential parts of a Laurent's polarimeter. It consists of two nicol prisms *P* and *A* mounted in brass tubes placed some distance apart and capable of rotation about a common axis, a half shade device *H*, so ingeniously invented by Laurent which divides the field of the polarised light emerging out of the nicol *P* into two parts generally of unequal brightness. A glass tube *D* with ground ends, against which fit plane glass plates, contains thė active solution, say, that of sugar. It is mounted between the two nicols . All the tubes are arranged in one line and are mounted on a rigid iron base.

Monochromatic light from a source S, rendered slightly convergent by a convex lens *L*, is made to fall upon the polarising nicol *P*. After passing through *P* the light is rendered plane-polarised with its vibrations in the principal plane of the nicol. This plane-polarised light before passing through the analysing nicol *A* is made to traverse the half-shade device *H* and the tube *D* containing the active solution.

The emergent light is viewed through a Galilean telescope *T*. The analysing nicol *A* can be rotated about its axis and its rotation can be accurately measured on a circular scale graduated in degrees with the help of vernier (V.S.) attached to it.

In some polarimeters the vernier scale is fixed and the circular graduated scale is capable of rotation with the nicol.

Half-shade device. It was to overcome the inability of the eye to judge the exact position of extinction, when the two nicols are adjusted in the crossed position that half-shade device had its origin. It consists of two semi-circular plates, one made of glass and the other of quartz. Both glass and quartz halves are cemented together. The quartz plate is cut parallel to its optic axis, which is parallel to YY', the line joining the two plates. It is a half-wave plate for sodium light, *i.e.*, its thickness is such that it introduces a path difference of half a wavelength between the ordinary (O) and extraordinary (E) waves in sodium light. In other words, in passing through the plate, ordinary wave *gains* half a wavelength on the extraordinary wave. The glass plate should be of such a thickness that it transmits the same amount of light as quartz plate does.

To understand the action of Laurent's plate let us suppose that the principal plane of the polarising nicol makes an angle θ with the optic axis of the quartz plate, *i.e.*, the principal plane of the polarising nicol is parallel to PP_1. When the ordinary monochromatic light after passing through the polarising nicol is incident normally on the half-shade plate the direction of vibrations of the plane-polarised light is parallel to PP_1. On passing through the glass half, the vibrations will remain along the same direction, *i.e.*, PP_1, because glass is not doubly refracting; but in the quartz-half, the vibrations will be split up into E and O components polarised at right angles to each other.

The B component is parallel to the optic axis of the quartz (XT') while the O component is perpendicular to it, *i.e.*, along XX'. These two rectangular components travel with unequal speeds within quartz and when they arrive at the opposite face, they differ in phase by π. Because of this phase difference of x the direction of the O component in quartz is reversed. Thus the two components OA and OC of the original incident vibrations along OP, on passing through quartz become OA and OD, the latter being the reverse of OC. These two component vibrations on emergence from quartz combine to form a linear resultant vibration, whose direction will be along OQ inclined with the optic axis OA (Y-axis) at the same angle θ as the incident vibration along OP with that axis but on the other side of it,

$$\text{i.e.,} \qquad \angle AOQ = \angle AOP.$$

Thus the effect of the quartz plate has been to rotate the plane of vibration by an angle 2θ.

Now if the principal plane of the A-nicol is parallel to P_1OP the plane-polarised light through the glass-half will pass unobstructed while that through the quartz-half is partially obstructed so that the glass-half will appear brighter than the quartz-half. On the other band, if the principal plane of the A-nicol is parallel *to* QQ_1 the quartz-half will appear brighter than the glass-half for the same reason as above.

It is this property of the half-shade device of dividing the field into two halves of unequal intensity that has made this polarimeter so successful.

When the principal plane of the A-nicol is parallel to OA (Y-axis)—a line which bisects the $\angle POQ$ – the emergent vibrations from both the glass-half and quartz-half are equally inclined to the principal plane of the A-nicol and hence the two components make the two halves *equally bright*.

When the principal plane of the A-nicol is parallel to OC (X - axis)—at right angles to the optic axis of the quartz – again the two halves are *equally bright* since OP and OQ are equally inclined to X-axis and the transmitted amplitudes OC and OD are equal. But the intensity in both parts of the field in this latter case being $I = I_0 \sin^2 \theta$, is small as compared with I_0 if $\theta \to 0$. This is the position of equal darkness in the two portions of the field. The variation in the intensity of the field as the A-nicol is rotated through a small angle about this position. A slight rotation of the -A-nicol from the setting say, in the clockwise direction causes a rapid increase in the transmitted amplitude *OD* from the quartz-half and a decrease in the transmitted amplitude *OC* from the glass-half. Thus, the quartz part appears brighter than the glass part. On the other hand, a slight rotation of the *A*-nicol in the anti-clockwise direction causes the glass part to appear brighter than the quartz part. It is in this setting of the analysing nicol that the eye can easily detect a slight variation in the intensity of light and the readings are therefore taken for this position. The advantage of this device over the extinction method lies in the fact that *the slightest rotation of the A-nicol from this critical position produces a considerable variation in the brightness of the two halves.*

The light coming out of the *A*-nicol is examined by a telescope. The position of the *A-nicol* is adjusted by rotating the tube containing, the *A*-nicol and the amount of rotation is measured with the help of the attached vernier scale.

The angle 2θ between *OP* and *OQ* is known as half-shadow angle and can be varied by rotating the *P*-nicol about its own axis. The sensitiveness of the apparatus depends upon the magnitude of this angle and there is certain *minimum* half-shadow angle for which this polarimeter is very sensitive.

It may be noted that this polarimeter is suitable for only one particular wavelength for which this half-shade device is made.

Example. *A 20 cm long tube containing sugar solution is placed between crossed nicols and is illuminated by light of wavelengh 6×10^{-5} cm. If the specific rotation is 60° and optical rotation produced is 12°, what is the strength of the solution ?*

The specific rotation *S* at a given temperature *t* and far a given wavelength of light λ used is given by

$$S_t\lambda = \frac{\theta}{I \times d},$$

where θ is the rotation, *d* is the density of the solution in g/cm^3 and *I* is its length in decimeters.

Hence $S = 60°$; $l = 2$ decimeters ; $\theta = 12°$

and $\lambda = 6 \times 10^{-5}$ cm.

$$\therefore \qquad \delta = \frac{\theta}{l \times S} = \frac{12}{2 \times 60} = \frac{1}{10}.$$

∴ it is a 10% solution of sugar, *i.e.*, 1 g of sugar is dissolved in 10 cm³ of solution.

Lippich Polarimeter

Laurent's half-shade device is suitable only for one particular wavelength for which the quartz half-wave plate is made. To overcome this difficulty other methods have been employed. Out of these *Lippich polarimeter* is one which needs a special mention.

In this case the polarise which consists of two nicols P_1 and P_2 with their principal planes inclined to each other at a small angle θ is placed in front of the aperture S_1. The principal plane of P_1 is parallel to OQ_1 and that of P_2 parallel to OQ_2 as indicated.

A is the analysing nicol placed behind another aperture S_2, and is capable of rotation about a horizontal axis and its position can be read on a circular scale attached to it. The light coming out of the *A*-nicol is examined by the Galilean telescope *T*. The tube *D* containing the optically active solution is placed between S_1 and S_2.

Incident light rendered parallel by a lens *L* is made to fall upon the polarising nicol P_1. Since one half of the field of view is illuminated by light passing through the nicol P_1 and the other half, by light passing through both the nicol P_1 and – P_2 the field looks divided into two halves of unequal brightness separated by a sharp-line which is the image of the lower edge of nicol P_2.

If the principal plane of the *A-nicol* is perpendicular to the principal plane of the polarising nicol P_1 (*i.e.*, OQ_1), one-half of the field of view is bright and if it is perpendicular to that of P_2 (*i.e.*, OQ_2) the other half is bright. On the other hand, if the analysing nicol *A* has its principal plane perpendicular to *OR* — the intermediate position—both the halves are equally bright, so the direction of *OR* sets the zero position of the *A*-nicol. It is evident that for greater sensitiveness of the apparatus, θ should have a small value. In fact the values of θ will depend upon the brightness of the source and the transparency of the solution under examination. On the whole it may be remembered that for very small values of θ, the intensity becomes too low.

Biquartz Polarimeter

A *biquartz* is another simple optical device which can be usefully employed for determining the rotation of the plane of polarisation. It consists of two semi-circular quartz plates *ACB* and *ADB* placed in juxta-position each about 3.75 mm thick, one of right-handed (*R*) and the other of left-handed (*L*) quartz, cut with their optic axes perpendicular to their refracting faces. They are cemented together to form a circular plate, this biquartz plate is placed just in front of the polarising nicol like Laurent's half-shade plate. When plane-polarised white light is passed through each half of this plate its different components will be rotated through different angles by each half but in the opposite sense. For sodium tight this rotation is about 90°. Thus if *QOP* be the direction of the incident vibration, it will be rotated to the right in the plate *ACB* and the vibration of the yellow component will be parallel to *OQ'*, *i.e.*, at right angles to *OP*. In the other plate this component is rotated through 90° to the left and its vibration will be parallel to *OF*. Hence the vibration of the yellow ray will be parallel to the same direction in both halves of the field *viz.*, to the line *P'OQ'* perpendicular to *POQ*. Thus, if the principal plane of the analysing nicol be parallel to *POQ*, sodium light will be extinguished in both halves of the field and the other colours will be present in the same proportion in each half so that they will exhibit the same tint, a grayish violet, called the tint of passage.

A slight rotation of the *A* -nicol one way or the other will change this tint either to blue or red. On passing through this position the transition from red to blue is very rapid. Thus the zero position can be obtained with a fair degree of accuracy. When this is secured the plane of polarisation of the incident light is perpendicular to the principal plane of the analyser. Thus this arrangement even though designed for yellow sodium light is quite sensitive with white light for the proper setting of the instrument.

When the tint of passage is attained the substance under investigation is placed between the biquartz and the analysing nicol. If the substance rotates the plane of polarisation, the tint of passage will disappear from the field of the analyser making one-half red and the other half blue and to restore it the nicol must be rotated in the direction in which the plane has been rotated and through an equal angle.

7

Radiations and Spectra

Sources of spectra. We are all familiar with Newton's experiment which proved conclusively that sunlight consisted of an infinite number of constituents all refracted by the prism at different angles but grouped together in seven different colours, red, orange, yellow, green blue, indigo and violet.

Besides solar light used by Newton, we have now studied the spectra emitted by number of sources of light, some of which we describe below:

The incandescent solid. A solid body heated to a high temperature has long been used as a source of light for the study of spectra. It emits a continuous spectrum and all the colours are gradually added up as the temperature rises higher and higher. A tungsten filament lamp is nowadays a common example of such a source of light. A modern efficient variety of this lamp is the coiled coil type filled with argon and nitrogen at reduced pressure to diminish evaporation losses. The temperature in this case is as high as 2100°C.

The Bun sen flame. Any substance introduced in the nonluminous part of the flame on being volatilized results in the characteristic spectrum which may consist of a number of spectral lines. To introduce the salt in the flame we use an asbestos ring dipped in the solution of the salt or we take a platinum wire, heat its one end and dip it in the salt which melts and adheres to it. It is then inserted in the flame. This is our usual method of obtaining a sodium flame.

The arc. When two adjustable carbon rods connected to electric supply mains are touched together and then slowly separated, brilliant arc is produced across the small air gap. A photographic examination of the arc reveals carbon are spectrum. When some volatile metal is inserted in the arc we will get the characteristic spectral lines of the metal. In place of carbon rods one may have metallic rods in which case the metal arc results. With metals of low melting point the arc may be maintained In an evacuated vessel. The most familiar examples of this are the mercury vapour lamp and the sodium vapour lamp.

Two forms of the mercury vapour lamp are in common use. The form (a) consists of an outer vessel A and an inner vessel B, the space between them being evacuated. The inner vessel contains a small quantity of mercury and argon ; sealed at the ends are two main tungsten spiral electrodes E_1 and E_2 with a starting electrode 5 connected to the distant electrode E_2 through a very high resistance In each of the main electrodes there is a rod of barium compound which emits electrons on applying A.C. voltage to them through a choke P which serves as a stabilizer. To start with argon between E_1 and S is ionised and later between E_2 and S so that a discharge starts warming up the interior which results in the mercury being gradually evaporated and the pressure within the lamp increases. In a few minutes the colour of the glow within the tube changes from the red of the neon to the greenish blue of mercury. The lamp in addition to intrinsic green violet gives a copious supply of ultraviolet light. The spectrum is found to consist of a number of lines which are used for the study of several phenomena.

The form (*b*) consists of an evacuated tube ending in two bulbs containing mercury. Two metallic electrodes are dipping in them and can be connected to the mains through a variable resistance *R*. By slightly tilting the tube an arc is struck which emits light in ultraviolet.

A laboratory form of a sodium vapour lamp. *A* is a U-shaped tube made of some special glass not blackened by sodium vapour. It has two oxide coated electrodes E_1 and E_2 sealed into its ends and contains sodium and neon. The electrodes are connected to the secondary of a transformer giving a sufficiently high voltage to start the discharge in the neon. The inner assembly is surrounded by a vacuum flask *F* to avoid heat losses and soon the temperature of *A* rises so that sodium volatilizes and carries the discharge, being more easily ionized than neon atoms. In a few minutes the working temperature of about 300°C is reached and the red colour of the neon gives place to intense yellow D_1 and D_2 sodium lines.

The spark. If two metal electrodes are connected to the terminals of the secondary of an induction coil or that of a high voltage transformer, a series of sparks may be produced across the gap provided it is small. The intensity of the spark can be increased by joining a condenser across the gap and the spark. This provides an intense source of radiation. The spectrum though characteristic of the metal forming the electrodes is different from that obtained by an arc. In the arc we have low voltages and high currents, whereas in the spark the voltages are as high as many thousands and the currents are small.

The vacuum or discharge tube. This is a common laboratory source for the spectral analysis of light emitted by a gas through which a discharge is passed at low pressure. Two simple forms of discharge tubes containing the gas under examination. It consists of two wider portions *A* and *B* joined together by a capillary tube C. The aluminium electrodes are fitted into the wider tubes and are held in position by platinum wires sealed into the tubes. By connecting the electrodes to an induction coil an electric discharge is passed through the gas in the tube and a brilliant luminosity results. Tubes of various shapes may be used to suit the convenience of the working conditions but the principle is the same.

Besides the sources mentioned above there are several others used in the present-day technique for the examination of spectra ; for instance, the fluorescent tube, the electronic bombardment and the electrodeless discharge obtained by means of Tesla coil which produce high frequency electrical currents If a partially exhausted tube without electrodes is placed inside the Tesla coil, it emits a brilliant glow which is characteristic of the gas at low pressures. Various other types of arcs such as pulsating direct current arc, the high voltage alternating current arc are also available for the purpose and are usefully employed in spectroscope.

Emission and absorption spectra. The spectra may be grouped into two principal classes known as emission spectra and absorption spectra. The former are due to the light emitted from any of the sources described in the preceding section, while the latter are due to the absorption of light energy by material media interposed in the path of light. This absorption may be *general* when the light of all wavelengths within the visible region of the spectrum is absorbed during its passage through a material medium or *selective* in which case some colours are absorbed and some transmitted such as in coloured glasses and dyes.

Emission spectra. It is usual to distinguish between two main types : (*i*) Continuous spectra, and (*ii*) discontinuous spectra.

Continuous spectra contains all the colours emerging from one to the other. They are emitted by any solid body when white hot. A dense luminous gas flame gives a continuous spectrum due to the presence of the incandescent carbon particles in it. The continuous spectrum is caused by the molecules in the solid being very close together and packed up so that collisions are frequently and constantly occurring and therefore all possible vibrations and wavelengths are emitted.

Discontinuous spectra may be classified into two groups: (*i*) line spectra, and (*ii*) band spectra.

(i) *Line spectra.* When light is emitted by a substance in the atomic state such as sodium vapour in a sodium lamp or a glowing gas and is examined by a prism spectrometer or grating, it is found to consist of a number of isolated bright lines of different wavelengths separated by dark gaps. Such a spectrum is called a line spectrum. Salts of sodium, potassium and calcium in the Bunsen flame give line spectra Similarly, if an electric discharge is passed through hydrogen or helium at low pressure, line spectra are produced. Such spectra are generally emitted by the substances in the *atomic* state and are *characteristic* of them. Very often these lines are regularly spaced and form what we call a *series.*

It may also happen that the lines may be quite irregularly spaced.

(ii) *Band spectra.* These spectra consist of a number of bright bands the intensity of which varies from one end to the other. In fact on examination by spectroscopes of high resolving power, it is found that each band consists of fine lines which are very close to each other at the sharp end of the band and are gradually more and more distant at the diffuse end. These spectra have a fluted appearance and are very often known as fluted spectra. Such spectra are emitted by the substances in the molecular state, when the heat is not too strong to break the molecules into constituent atoms. The carbon compounds mostly reveal this type of spectrum. In the discontinuous spectrum, the molecules and the atoms are on the average much farther apart and do not interfere with each other's radiation.

Although, in general, line spectra originate in atoms only and band spectra in molecules, yet the same substance may give a variety of spectra when excited to luminosity by different means. This is the case with several substances such as sodium, nitrogen etc. With sodium in Bunsen flame two close yellow lines are observed while the photograph of sodium arc reveals many spectral lines. Again if the spark of sodium is examined by a spectroscope certain other lines are found to be present. We then have three different kinds of spectra, all characteristic of the sodium with some common lines but no two spectra are exactly alike.

Spectrum analysis. The work of Bunsen and Kirch off revealed many interesting facts about spectra, most important being that the emission spectra particularly in the vaporous condition depend entirely on the nature of the substance. Therefore, the spectroscope examination of the light emitted by a substance enables us to find its composition. The method is *rapid* and *sensitive,* for even a small trace of the substance in the Bunsen flame is sufficient to give the characteristic lines of the substance. This method of identifying a substance is called spectrum analysis. The spectra of all substances have been studied and tabulated so that if we want to find out the composition of an unknown substance the arc spectrum of the substance is examined in a spectroscope and compared with the known spectrum. This method has made possible the discovery of a number of elements available in small quantities such as caesium, rubidium, thallium, helium, etc.

Besides this practical application of the spectra of the elements, a good deal of information is obtained from them with regard to the structure of the atoms. The spectra are in fact the language of the atoms which has been interpreted now and most of our present information about atom building has been obtained from their study.

Spectra of various substances. We have now studied spectroscopically the spectra of various elements and find that they differ greatly in complexity. Thus structure of the *H*-spectrum is the simplest as it gives four bright lines, in the visible region, one in the extreme red, one in the blue, one in the blue-violet and one in the violet. These lines are known as

$$H_\alpha, H_\beta, H_\gamma, \text{ and } H_\delta.$$

Besides we know now, that the hydrogen spectrum extends in the ultra-violet and infra-red regions giving several other series of lines. The next in the order of simplicity are the spectra of the alkali metals, lithium, sodium, potassium, etc. Their spectra are strikingly similar to each other. We find that most of the lines are doublets as in the case of sodium where we have D_1 and D_2 lines very close altogether with wavelengths 5896 and 5890 Å.

By means of the spectroscope the spectra of most of the elements have been investigated and in the case of rare earths three close lines have been discovered. The spectra of some substances are very complex. Iron, for instance, gives as many as 5300 lines which have all been measured accurately.

The spectra of elements somewhere in the middle of the periodic table contain in general a large number of lines apparently distributed at random but in most cases they are regularly spaced and arranged in some systematic order.

Hydrogen spectrum. The order in the spectral lines of hydrogen was first discovered by Balmer in the year 1885, when he was investigating its spectrum. He found, as a result of close study, that the frequencies of four lines in the visible region together with several more in the near ultra-violet could be represented by the simple formula

$$v = R\left(\frac{1}{2^2} - \frac{1}{n^2}\right) Hz, \qquad ...(1)$$

where $n = 3, 4, 5,$.......for various lines and R is a constant having the value **3.291 × 10¹⁵ s⁻¹**. It is now called Rydberg's constant, named after a Swedish spectroscopes.

If we put $n = 3$, we get H_α line, for n=4, H_β line and so on.

It is convenient to express the above formula in terms of the wavelength.

Now $\quad v = \dfrac{c}{\lambda},$

$$\therefore \quad \frac{c}{\lambda} = 3.291 \times 10^{15}\ s^{-1}\left(\frac{1}{2^2} - \frac{1}{n^2}\right) \quad \text{... by (1) above}$$

or
$$\frac{1}{\lambda} = \frac{3.291 \times 10^{15}\ s^{-1}}{3 \times 10^8\ m\ s^{-1}}\left(\frac{1}{2^2} - \frac{1}{n^2}\right)$$

... [∵ c, the velocity of light = 3×10^8 ms⁻¹

$$= 1.09678\ m^{-1} \times 10^7\left(\frac{1}{4} - \frac{1}{n^2}\right) \qquad ...(2)$$

The ratio $1/\lambda$ or the reciprocal of wavelength is called the wave number, that is, the number of waves per m in vacuum and now the Rydberg constant has the value 1.09678×10^7 m⁻¹. The wave number is denoted by $\bar{v}$.

Simplifying further, we have

$$\bar{v} = \frac{1}{\lambda} = 1.09678 \times 10^7\ m^{-1}\left(\frac{n^2 - 4}{4n^2}\right)$$

or
$$\lambda = \frac{4}{1.09678 \times 10^7} \cdot \frac{n^2}{n^2 - 4}\ m$$

$$= \frac{4 \times 10^{10}}{1.09678 \times 10^7} \cdot \frac{n^2}{n^2 - 4}\ Å. \quad [\because 1\ m = 10^{10}\ Å]$$

$$= 3645\ \frac{n^2}{n^2 - 4}\ Å, \quad \text{where } n = 3, 4, 5, 6, \text{etc.}$$

This expression gives the wavelengths of the Balmer series in hydrogen in excellent agreement with the observed values.

We find that the lines get closer together as we go towards the shorter wavelength side crowding up to a series limit so that by putting $n = \infty$ in the formula

$$v = R\left(\frac{1}{2^2} - \frac{1}{n^2}\right),$$

we have for the frequency of the shortest line

$$v = \frac{R}{4} \text{ in the visible region.}$$

This work of Balmer was extended later by other experimenters and several other spectral series have been discovered.

(*i*) *Lyman series* in the ultra-violet,

$$v = R\left(\frac{1}{2^2} - \frac{1}{n^2}\right), \text{ where } n = 2, 3, 4 \text{ etc.} \quad \ldots (3)$$

The lines crowd to a series limit at $v = R$.

(*ii*) *Ritz Paschen series* in the infra-red,

$$v = R\left(\frac{1}{3^2} - \frac{1}{n^2}\right), \text{ where } n = 4, 5, 6 \text{ etc.} \quad \ldots (4)$$

(*iii*) **Brackett series** in the far infra-red,

$$v = R\left(\frac{1}{4^2} - \frac{1}{n^2}\right), \text{ where } n = 5, 6, 7 \text{ etc.} \quad \ldots (5)$$

(*iv*) *Pfund series* in the extreme infra-red,

$$v = R\left(\frac{1}{5^2} - \frac{1}{n^2}\right) \text{ where } n = 6, 7, 8 \text{ etc.} \quad \ldots (6)$$

More generally, all these series can be denoted by the single equation,

$$v = R\left(\frac{1}{n_1^2} - \frac{1}{n_2^2}\right) \quad \ldots (7)$$

where n_1 and n_2 are integral numbers. For Lyman series $n_1 = 1$. For Balmer $n_1 = 2$ and for Paschen, Brackett and Pfund series $n_1 = 3$, 4 and 5 respectively.

Thus we find that the frequency of a given line can be expressed as a difference of two spectral terms represented by

$$R, \quad \frac{R}{4}, \frac{R}{9}, \frac{R}{16} \ldots\ldots \frac{R}{n^2}.$$

These terms it appears represent the energy states or energy levels in the mechanism of radiation from an atom.

Energy levels. A number of *energy levels* are represented by a set of horizontal lines such that their vertical distances below the line OO' are proportional to the values of the terms

$$R, \quad \frac{R}{4}, \frac{R}{9}, \frac{R}{16}, \ldots \ldots$$

Transition between any two of these energy levels gives a corresponding line in the spectrum.

Thus the frequency of the first line of Lyman series is given by the difference between R and $R/4$ levels and that of the second by the difference between R and $R/9$, that is, all the lines of the series are represented by the vertical separation between the lowest level and any of the upper levels. Similarly, for Balmer series the lines are represented by the vertical separation between the $R/4$ energy level and any one of the upper level and so on, for other series. Of course with the complex spectra the picture is not so simple as in the case of hydrogen but the analysis of the lines has shown that it is possible to represent all the lines of the spectrum as difference between different members of a set of terms. For explanation of hydrogen spectrum of Bohr's theory.

Absorption spectra. If light from an arc lamp giving a *continuous* spectrum is examined by a spectroscope after it has passed through a gas at low pressure and lower temperature than that of the source it is found that the spectrum has certain dark regions, showing that certain wavelengths are absorbed. This is called absorption spectrum of the substance. As in the case of emission spectra, we have both the *continuous* or *general absorption* and *selective* or *line absorption.*

If red or blue glass is placed in front of the sunlight or arc light, it only allows a narrow band of red or blue rays to be transmitted giving a continuous absorption spectrum. In the case of cobalt glass, the spectrum is crossed by three dark bands indicating that it absorbs three different groups of waves. There are many substances, solids and liquids, such as potassium salts and aniline dyes which produce such spectra giving rise to isolated dark bands whose number, width and position are characteristic of the absorbing substance. It is possible, therefore, to use this property in the analysis of the solutions of such substances just like the emission spectra. A highly developed technique of absorption spectroscopy is made use of for the identification of dyes, drugs, oils, glasses, etc. On the other hand, in the case of an absorbing gas or vapour narrow sharp lines have been observed in absorption spectra as in the emission spectra, that is, absorption occurs for certain definite isolated wavelengths. This can be demonstrated by a simple experiment.

Suppose a sodium flame is interposed between the arc lamp and the slit of the spectrometer it is found that the spectrum is crossed by two dark lines in the yellow portion. These dark lines occur exactly in the same place as is normally occupied by the bright sodium lines, a fact which can be verified by holding an opaque screen in front of *S*, so that the spectrometer slit is illuminated entirely by the light from the sodium flames. This means that sodium vapour absorbs those wavelengths which it itself is capable of emitting when acting as a luminous source. This is the absorption spectrum of the sodium lines and the phenomenon is known as the Reversal of Sodium Lines. Theoretically the phenomenon can be explained as a case of resonance as the ultimate particles absorb energy from the continuous spectrum corresponding to their own natural frequencies. Kirchhoff and Bun sen as a result of their study of absorption spectra formulated a law concerning the emission of light by substances known as Kirchhoff's law.

It states that—a substance which emits a light of certain wavelength when heated to incandescence will absorb the light of the same colour when cold if light from a source at a higher temperature is made to pass through it. It is the light corresponding to yellow lines in the arc light (higher temperature) which is absorbed by the sodium vapour in the flame (lower temperature) that we get dark lines in their place In fact these lines are not dark but comparatively less luminous than the rest of the spectrum.

Solar spectrum and Fraunhofer lines. When we examine solar spectrum carefully by a spectrometer, we find that it is crossed by a large number of narrow dark lines distributed throughout the whole length of the spectrum. They were first observed by Wollaston in 1802 and later studied by Fraunhofer, a German optician, in 1814, and, therefore, called after him as Fraunhofer lines.

Fraunhofer counted some 600 of these lines and measured the positions of 324 and showed that they occupied definite and constant positions in the solar spectrum. He denoted the chief lines by letters *A, B, C, D, E, F, G,* and *H,* the lines *D* corresponding to sodium, C and *F* due to hydrogen and so on. The interpretation of these lines remained a mystery until the work of Bunsen and Kircchoff on spectra appeared in 1859.

According to them these lines are produced by absorption in the atmosphere of the sun. The central part of the sun called the *photosphere* is intensely hot and emits a continuous spectrum. Surrounding this central part there is an atmosphere of vapour and gases called the *chromosphere* which is comparatively colder. When light coming from the interior of the sun passes through the chromosphere, its vapours absorb light of certain wavelengths which they themselves would emit, if incandescent. Thus if the vapours of sodium, iron and copper are present in the chromosphere, white light in passing through them would lose those colours (wavelengths) which these elements would emit at the higher temperature. The corresponding wavelengths would, therefore, be missing when the solar spectrum is examined. By making a comparison of missing wavelengths with the wavelengths emitted by various elements, it is possible to discover the elements composing the sun. The relative intensity of the bright and dark lines shows the relative abundance of the elements in the chromosphere. It has been found that the sun's atmosphere consists of vapours of as many as 60 elements existing on the earth.

It may be noted that some of these lines are also due to absorption by oxygen, ozone, water vapour and carbon dioxide in the earth's atmosphere.

A table of chief Fraunhofer lines is given below in which the wavelengths are expressed in terms of Angstroms (Å).

$$1\ \text{Å} = 10^{-8}\ \text{cm}.$$

Sometimes other units are also used in the measurement of wavelengths.

$$\text{A micron } (1\ \mu) = \frac{1}{1000}\ \text{mm} = 10^{-3}\ \text{cm} = 10^{-6}\ \text{m}$$

$$\text{A milli-micron } (1\text{m}\mu) = \frac{1}{1000}\ \text{micron} = 10^{-6}\ \text{cm} = 10^{-7}\ \text{m}$$

$$= 10^{-9}\ \text{m}$$

$$1\ \text{m}\mu = 10\ \text{angstroms}$$

$$\text{An Angstrom (Å)} = 10^{-8}\ \text{cm} = 10^{-10}\ \text{m}$$

$$= 10^{-7}\ \text{mm} = 0.1\ \text{m}\mu.$$

Spectra of heavenly bodies. We have just discussed the solar spectrum. An examination of the spectrum of the moon reveals that it is identical with that of the sun. This is due to the fact that moon has no atmosphere of its own which could rob the sun-light reflected from it of any of the frequencies present in it. In the case of most of the other planets absorption bands are found to exist in the light they reflect. Thus carbon dioxide has been shown to occur in the atmosphere of Venus, ammonia and methane in the atmosphere of Jupiter and latter alone in that of Mars. It is also possible to estimate the amounts of these compounds from the intensities of the absorption bands in their respective *spectra.* The stellar spectra have also been photographed and they show the presence of the same elements as are found on the earth. Iron, calcium, hydrogen and titanium are prominent amongst them. The study of the stellar spectra has been useful in several other ways. We have been able to find the distances, brightness, surface temperature, radial velocity, size and mass of the stars. It is from the type of the absorbing spectrum that we can estimate the temperature of the absorbing material. For instance, by examining the relative intensities of the absorption lines of hydrogen, calcium and iron in the spectrum of *Arcturus,* the surface temperature is estimated at about 4000°C.

Table showing wavelengths of Fraunhofer lines in Å

Line	*Element*	*Wavelength*	*Line*	*Element*	*Wavelength*
A	Oxygen	7661	*h*	$H_{(\delta)}$	4102
B	Oxygen	6867	*H*	C_a	3968
C	$H_{(\alpha)}$	6563	*K*	C_a	3934
		5896			
D	*Na*	5890	. .	$\mu_g - C$	3838
E	E_s	5270	*L*	$E_s - C$	3820
		5178			
b	μ_g	5173	*M*	F_s	3720
F	$H_{(\beta)}$	4861	*N*	F_s	3581
f	$H_{(\gamma)}$	4340	*O*	F_s	3441
G	F_s	4308	..	$T_j - F_s$	3057

The Doppler effect in light. We know from our every day experience in the case of sonorous bodies that if there is relative motion between the source and the observer, an apparent change in pitch takes place. This is called Doppler effect. When the source approaches the observer, there is an apparent rise in pitch and hence a decrease in wavelength and *vice versa*. The same phenomenon is observable in light provided the relative velocity v of the stellar source and the earth along the line of sight is appreciable in comparison with the velocity of light c. The wavelength of every spectral line undergoes a change. Suppose the source is in motion. Let f be the frequency of the source, v its velocity towards the observer and c the velocity of light. Since the waves produced per second are crowded into the space $(c - v)$ when the source approaches the observer, the wavelength of light will be shortened from

$$\lambda = \frac{c}{f} \text{ to } \lambda' = \frac{c-v}{f}.$$

The observed frequency of the light $f' = \frac{c}{\lambda}$, then

$$f' = f\,\frac{c}{c-v}. \qquad \text{... (8)}$$

If the source is receding, the sign of v is reversed with the result that

$$f' = f\,\frac{c}{c+v}. \qquad \text{... (9)}$$

Combining these results we have the apparent frequency

$$f' = \frac{c}{\lambda} = \frac{f}{\left(\frac{c \mp v}{c}\right)} = \frac{f}{\left(1 \mp \frac{v}{c}\right)} = f\left(1 \mp \frac{v}{c}\right)^{-1} \qquad \text{... (10)}$$

If the observer is in *motion* with velocity v then during approach it will actually receive more waves per second equal to v/λ and during recession the number of waves received will be reduced by v/λ, therefore, the apparent frequency,

$$f = f \pm \frac{v}{\lambda} = f \pm \frac{v}{c}\,f = f\left(1 \pm \frac{v}{c}\right). \qquad \text{... (11)}$$

It should be noted that in light there is no physical difference between the two cases whether the source is in motion or the observer is in motion because this distinction is based on the assumption that there is a medium, such as air, for the propagation of sound waves while no such medium has been discovered for light waves (Relativity theory).

Expanding the right-hand side of equation (10) by the binomial theorem, we have

$$f = f\left(1 \pm \frac{v}{c} \pm \frac{v^2}{c^2} \pm \frac{v^2}{c^2} + \ldots\ldots\right),$$

which is the same as equation (11), if v^2/c^2 and higher terms are neglected.

So in actual practice, equation (11) is used.

If λ' is the altered wavelength and λ the actual wavelength, the equation (11) becomes

$$\frac{\lambda'}{\lambda} = 1 \mp \frac{v}{c} \quad \text{or} \quad \frac{d\lambda}{\lambda} = \mp \frac{v}{c}. \qquad \ldots(12)$$

This change in wavelength can be observed when the stellar spectra are photographed alongside that of an arc or spark giving some of the same lines as in the stellar spectrum.

If the corresponding stellar lines are shifted towards the violet end indicating a diminution in wavelength, we conclude that the star is approaching us, and if the shift is towards the red end indicating an increase in wavelength, the star is receding from us.

Hoygens detected the displacement of the *F*-line towards the red end in the spectrum of Sirius as compared with the spectrum of sun, indicating a recession of 48 km/sec. Similar observations have indicated that the stars Sirius, Rigel, Regulus and Astor are receding from the Earth while Arcturus and Vega are approaching the earth. Most of the stars have been observed to have radial velocities ranging from 10 to 300 km/sec.

There are a few other applications of the Doppler effect :

(i) *Angular velocities of the sun and the planets.* By comparing the position of the Fraunhofer lines in the spectrum of two different parts of the surface of a planet at a time we may find different shifts in the two cases indicating that all the parts of the planet do not have the same velocity towards the earth. By measuring this shift angular velocities may be determined. By this method we have discovered that when the spectrum of the west edge of the sun is compared with that of the east edge the absorption lines show a displacement corresponding to a velocity difference of about 2 kilometers/sec. A similar comparison of the spectra from the north and south edges of the sun shows no such displacements. Thus we conclude that the sun rotates about an axis passing through north and south.

(ii) *Discovery of double stars.* Some stars which ordinarily appear single are in fact double stars known as *Spectroscope binaries.*

They revolve about each other so that when one is approaching the earth, the other is receding from it. Their spectrum on one night may show all the lines as single. This would happen when one star is covered by the other and there is no component of velocity in the line of sight.

When both the stars become visible after a few nights on account of the approach of one and the recession of the other, the displacement of the spectral lines from the two will be in opposite directions and double lines would result. By this method we have discovered that several stars are double.

(iii) *The red shift.* The most important application of the Doppler effect is that the distant spiral nebula is found moving away with velocities as great as 14300 miles/ sec. Bright lines in the spectra of these island universes outside our own Galaxy (the Milky Way) appear to be displaced towards the red end of the spectrum, an effect known as the *red shift.* These observations have given rise to the belief that our universe is expanding.

(iv) *Saturn rings.* The telescope shows three concentric rings round the planet Saturn. It has long been disputed whether these rings were solid or composed of a swarm of smaller satellites In the former case the velocity of the outer edge of a ring would be greater than that of the inner one whereas in the latter case, the satellites nearer to the Saturn would have greater velocity than those farther away to balance greater gravitational pull.

[Since for Satellites motion,

$$\frac{mv^2}{R} = \frac{GMm}{R^2} \quad \therefore \quad v^2 = \frac{GM}{R}$$

that is, v increases as R decreases.]

This means that the inner-most edges of a ring would show a larger Doppler shift than its outer-most edges. This result has been confirmed by actual measurements of the shift. Therefore, it is concluded that rings are composed of swarms of individual satellites.

The invisible radiations. We have considered above that portion of the spectrum which when incident on the eye causes sensation of sight. But extending on either side of this visible portion there are invisible radiations which are too long or too short to affect the eye but they carry energy and obey the same laws as light, can be polarised and diffracted by concave reflection gratings. The portion beyond the red consists of longer wares extending upto 400,000 Å and is called the infra-red and that beyond violet consisting of shorter waves extending upto 136 A is called the ultra-violet.

The infra-red. This part of the spectrum was first discovered in 1800 by Sir William Herschel who observed that when the blackened bulb of a thermometer was placed even at some distance beyond the red end of the spectrum, it indicated a rise in temperature. This radiation is emitted copiously by the sun and other hot bodies and contains most of the heat energy of the sun rays. It is generally detected by the *heating effect* for which we may use the blackened bulb of a thermometer or a Crooke's radiometer. But more sensitive instruments like thermopile, Langley's bolometer and Boy's radio-micrometer are now used. *(For the principles underlying the working of these instruments and their use the student is advised to consult some books on Electricity and Heat).* As a result of the study of these rays we find that the maximum heating effect is always beyond the red end of the spectrum in the case of electric lamp and electric arc and in the visible part if the sun is the source. The spectrum of this part is continuous like the visible part and contains a number of dark lines where the heating effect is nil. Most of the properties of these rays are common with the visible portion and these rays extend so far as to merge with the radio waves at about 4×10^{-2} cm.

Since 1880 the infra-red portion of the spectrum has been carefully investigated and the wavelengths of lines constituting this portion have been measured. In the study of this region, we make use of a spectrometer with prisms and lenses made of *rock-salt* and *sylvine* for different ranges of wavelengths because glass absorbs them. It is also better to dispense with lenses altogether and to collimate the beam incident on the prism and to focus the emergent beam, stainless steel mirrors are employed.

The various methods that have been used for detecting and investigating infra-red rays are : (*i*) *Heating effects*, (*ii*) *Photography with special plates,* and *(iii) Photoelectric effect.* We shall now discuss these methods and the uses of these radiations.

(i) *Heating effects—Infra-red ray spectrometer.* To investigate the intra-red rays spectrum, we make use of the heating effects of the rays. A special type of infra-red ray spectrometer is used for examining the intensity of these rays and for determining their wavelengths. A powerful source of infra-red radiations illuminates the slit S_1. The radiation after reflection from the mirror C_1 is incident on a rocksalt prism P and after refraction through it the emergent beam is allowed to fall on a plane mirror M (both P and M are mounted on the rotating table). This beam after reflection from M falls on a concave mirror C and is brought focus on the detector T, which may be a *thermopile* or a *bolometer* connected to a sensitive galvanometer.

By keeping the detector fixed in position the prism table is rotated to bring the different lines in the infra-red spectrum on to the detector. This is indicated by the deflection of the galvanometer connected to the thermopile which rises to maximum when the line falls on the slit S_2. Thus the spectrum of any source can be mapped out. To actually measure the wavelengths of any source, we have first *to calibrate* the instrument by taking a source of infrared radiation whose wavelengths have already been measured by a reflection grating. The readings of the prism table are taken for each known wavelength and are plotted as abscissa and the corresponding readings of the galvanometer as ordinates.

(ii) *Infra-red ray photography audits uses.* Photography is another means of detecting infra-red radiation and is utilised with advantage in the petrography of the rays. The serious drawback of concave reflection grating spectra is that only a small portion of the incident radiation goes into the diffracted spectra which are therefore weak. Consequently exposures extending over several hours are necessary to get the spectral lines bright enough to be measured accurately. This is done with special photographic plates.

The infra-red ray photography has been very usefully employed for taking the photographs of distant landscapes in the dark even, because the rays can penetrate easily through the atmosphere. Dense fog and smoke are transparent to them. For this purpose special panchromatic photographic plates and infra-red ray filters are employed. A solution of iodine in alcohol is a suitable filter. It allows to pass through infra-red rays easily but is opaque to visible light.

In World War II infra-red ray photography was successfully used in revealing the positions of camouflaged buildings, in seeing through fog, in detecting enemy concentrations in the dark.

With the manufacture of special kinds of glass which is transparent only to a particular region of the spectrum the infra-red rays have been successfully employed for *secret signalling*. This requires a powerful source of infra-red radiation, which is so hot that it also emits light. But the light radiation which will spoil the secrecy, is cut off by the special glass screen transparent only to the infra-red.

(iii) *Photoelectric effect of infra-red rays.* This effect relates to the emission of electrons when radiations are incident on the surface of an alkali metal and will be discussed in detail. It may be usually applied to study the infra-red rays if a photo-electric cell of some suitable material such as rubidium is employed.

On account of their photo-electric effect these radiations are used in television for illuminating the objects. They are more suitable than ultra-violet radiations since the latter have harmful effect on the persons being televised.

In addition to the above applications due to their easy absorption by human body they have been used for *therapeutic purposes* such as the diagnosis of superficial tumours and varicose veins, in, curing infantile paralysis, sprains, dislocation of bones, etc.

The ultra-violet. Beyond the violet end of the spectrum, there is the ultra-violet portion. It spreads over several octaves ending in X-rays and then in ϒ-rays. It extends right up from 4000 A to 136 A. This region was discovered by Hitter in 1801 while investigating the photographic action of ordinary light on silver chloride. He found that the effect did not stop at the violet end of the spectrum but extended far beyond it.

The ultra-violet radiation is reflected in the same way as visible light but some metals behave in an interesting manner towards it. A thin layer of silver, for instance, reflects 95% of the incident light in the visible part of the spectrum but only 4% in the ultra-violet, thus forming a very good filter for transmitting the ultra violet while intercepting the visible. Nickel, magnesium and many other alloys act in a similar way and even more effectively than silver.

In 1811, Thomas Young formed Newton's rings with ultra-violet radiation and photographed them. He found that the radii of the corresponding rings were smaller with ultra-violet light than with visible light, thus showing that their wavelength was smaller than that of visible light. Subsequent experiments with reflection gratings confirmed that ultra-violet light consisted of waves in ether differing from light only by its smaller wavelength.

The ultra-violet solar spectrum is found to be continuous but traversed by a number of dark lines like the visible spectrum. These radiations obey the usual optical laws, can be caused to interfere and can be polarised by tourmaline crystals just like visible light. Besides the sun which is the most powerful source of ultra-violet radiation, we have several artificial sources rich in ultra-violet such as are lamps with iron or carbon electrodes, mercury vapour lamps, the spark spectrum, the mercury are aclosed in a quartz tube and discharge tubes.

The chief methods for detecting and investigating ultra-violet radiations are (*i*) *Photography,* (*ii*) *Fluorescence,* (*iii*) *Phosphorescence* and (*iv*) *Photo-electric effect.*

(i) *Ultra-violet ray photography.* To measure the wavelength of these rays and to investigate other properties and effects, we employ the photographic action of the rays. Since these radiations are easily absorbed by glass, the spectrographs make use of quartz and fluorspar prisms and lenses, the latter being transparent upto 1200 Å.

Concave reflection gratings have been employed for measuring their wavelengths as in the case of infra-red radiation but as the diffraction spectra are very weak on account of the poor reflecting capacities of the metals for ultra-violet radiations and also the energy being spread over several spectra, prism spectrometers are generally preferred. They give comparatively much intense spectra and combined with long photographic exposures, very satisfactory results are obtained.

(a) *Ultra-violet ray spectrograph.* Since quartz of which the prisms and lenses are made is doubly refracting it results in doubling of each spectral line and thus decreasing its sharpness and the resolving power of the spectrometer. To eliminate this defect Cornu prism is used which is made by holding together with a thin film of glycerine two similar right angle 30° prisms each cut with its optic-axis parallel to the base. One of them is of *right-handed* and other of *left-handed* quartz so that double refractions balance and the sharp spectrum results. The collimating lens and the telescope objective are respectively made of right-handed and left-handed quartz and are cut with the axes parallel to the optic axis of the quartz.

In order that a wide range of wavelengths may be focused on the photographic plate *G* to cover up the differences in the dispersive power it is tilted so as to be at about 20° to the axis of the lens.

(b) *Littrow spectrograph.* A compact form of the spectrograph for precision spectrography is the one designed by Littrow and is known as Littrow spectrograph. Light from the source under examination is passed through a slit *S* and is totally reflected by a quartz prism *R*. It then fails on a lens *L* and emerges as a parallel beam to become incident on a quartz prism *Q*, whose back surface is suitably coated to make it reflecting for ultra-violet light. The prism *Q* is set in the minimum deviation position for the sodium light and the refracted beam strikes almost normally at the back surface, which reflects the light back through the prism. The dispersed beam then passes through the lens which forms an image of the spectrum on the photographic plate *P*. Different regions of the spectrum are brought to focus on the plate by rotating the prism *Q* which it mounted on a rotatable table The prism *R* is arranged in such a way that the returning rays *pass over* it to the plate *P*. To absorb reflections from the lens surface to avoid fogging of the image, a small blackened screen is placed behind *L* and *R*- The whole instrument is enclosed in a light-tight box mounted on a straight girder.

This spectrograph has the additional advantage that the optic axis of the quartz being normal to the back surface of the prism, no double refraction occurs and any splitting due to the resolution of light into two opposite circularly polarised components is exactly compensated when the ray retraces its path out of the prism again. Such exact compensation, however, is not possible with Cornu prism on account of the difficulty of preparing two prisms of identical dimensions, one of left-handed and the other of right-handed quartz. Littrow spectrograph has also been successfully employed with a plane reflection grating for the study of very short and very long wavelengths in an optical spectrum.

Various other types of spectrographs and mountings have been employed depending upon the region to be investigated as the materials used for transmission have different powers of absorption for different regions. For wavelengths lying below 1850 Å vacuum spectrographs with concave gratings have been found most suitable. With such arrangements in the years 1890-1900. Schumann investigated upto 1200 Å. This is known as the Schumann region of the spectrum. Lyman, Millikan, Seighbahn and others have photographed lines of as short wavelengths as 200 Å.

(ii) *Fluorescence.* When ultra-violet light is incident on certain substances, they emit visible light. For example, when a solution of quinine sulphate in dilute sulphuric acid is exposed to the ultra-violet portion of the spectrum it is found to glow brightly emitting blue light. This phenomenon is termed fluorescence. It is the process in which a substance absorbs a part of the incident ultra-violet light and then immediately re-emits an appreciable part of it with its wavelengths longer than those absorbed. This act is known as Stokes' law. The energy of the absorbed light raises the atoms or the molecules of the fluorescent substance from the lower to the higher energy levels and when these atoms or molecules return to their normal states light of longer wavelength is emitted and in accordance with quantum theory, the loss in energy or the energy absorbed by the fluorescent substance is $h(v_a - v_e)$, where h is the Planck's constant and e_a and v_e are the frequencies of the light absorbed and emitted respectively. This is also what Stokes' law predicts. There are, however, cases where Stokes' law is violated because sometimes the energy of the excited atoms and molecules of the material are added to the energy of the emitted waves, the atoms and the molecules themselves returning to the normal or some lower energy level (*see* Raman effect).

There is a large number of substances which exhibit fluorescence such as uranium glass, gems, eocein, kerosene oil and various lubricating oils. Paraffin oil when exposed to sunlight exhibits a blue fluorescence. An aniline derivative termed fluorescein emits a yellowish green light when exposed to day light or an electric arc.

In the commercial fluorescent lamp the inner side of the tubular bulb is coated with fluorescent material called *phosphors* which absorbs the invisible ultra-violet light from the low pressure mercury arc and emits visible light. Fluorescent screens of cathode ray oscilloscopes and television tubes and to make visible with *X*-rays the shadows of bones, are other important applications of the phenomenon of fluorescence.

It may be noted that fluorescence persists so long as the fluorescent substances remain exposed to incident ultra-violet light and re-emission of light ceases as soon as incident light is cut off.

(iii) *Phosphorescence.* There are some materials in which the molecules excited by the absorption of incident ultra-violet light do not immediately return to their original state and the emission of light continues even after the exciting radiation is removed. This type of *delayed* fluorescence is called phosphorescence. The phenomenon is caused chiefly by the ultra-violet and violet parts of the spectrum and is usually accompanied by fluorescence, which may be used as a method of detecting and investigating the ultraviolet region. The common substances which exhibit this effect permanently are sulphides of calcium, barium and strontium. Luminous paints which are composed of these sulphides combine to phosphoresce for several hours in a dark room after exposure to sunlight. The coating materials used in fluorescent tubes show this effect to some extent, as we often notice that the glow is still faintly visible for some time after the electric current is switched off.

There is, however, no *essential* difference between fluorescence and phosphorescence except that the latter persists after the exciting radiation has ceased, whereas in the former, light is re-emitted within 10^{-7} or 10^{-8} s. There are gases, liquid as well as solids such as soda glass etc. which, when bombarded by charged particles such as electrons, emit light. They are also called fluorescent substances. The emission process is the same but the excitation of the atoms of the substance results from the electric field of the charged particles rather than from the electric field of the light waves.

(iv) *Photo-electric effect of ultra-violet rays.* The ultra-violet radiations are most effective in ejecting electrons from alkali metals which thus acquire a *positive* charge on being exposed to ultra-violet radiations so that if a plate of such a metal is joined to a quadrant electrometer or a sensitive electroscope a deflection will occur. If now a sheet of glass is interposed between the metal plate and the source of radiation, the effect may be stopped in most cases thus showing that it is due to the ultra-violet rays which are now absorbed by the glass plate.

Applications of Ultra-violet Radiations

(1) The ultra-violet radiations within the range 2000 Å and 3000 Å are very active biologically. They kill bacteria and are very effectively used for sterilizing air, water supplies and other materials in bulk.

(2) *Ultra-violet ray therapy* is now a medical technique of curing virulent ulcers, some skin affections and infected wounds. The rays are found to be an essential element in the manufacture of vitamin *D* in the skin, so necessary to healthy life. It is for this reason that persons suffering from rickets, tubular abscesses and diseases of the bone are advised regular exposures to sunlight or artificial sources of ultra-violet rays. Miners who are deprived of the normal amount of ultra-violet radiation which come from the sun are given occasionally artificial dose of these radiations to partly make up the deficiency caused by underground living. They assist in increasing the rate at which calcium and other minerals art absorbed in the system, stimulate the nervous system and ductless glands and help oxidation of fats.

(3) The difference in the fluorescence caused by them in different types of inks enables us to detect *forged* documents. For the same reason we can distinguish between real and false gems, artificial and real teeth.

(4) Fluorescent lamps depend for their action upon the fluorescence caused by the ultra-violet light in phosphors coated inside the fluorescent tubes.

(5) We have already learnt that the resolving powers of the telescope and the microscope depend upon the wavelength of the light used, the smaller the wavelength, the greater the resolving power. So the resolving power increases if the visible light is substituted by ultra-violet light. The grain of the image improves and more details are observed Ultra-violet radiations have enabled us to photograph bacteria even.

The complete radiation spectrum. In our study of the optics we have dealt with the visible, the infra-red and the ultra-violet portions only. Work during the last several decades has resulted in the discovery of radiations that are hundred million times longer than the red known as radio waves, others given off in the *sub-atomic* transformation are ten million times shorter than the violet, all essentially resembling light in their physical nature. They all travel with the velocity of light equal to $3 \times 10^8 \text{ ms}^{-1}$ in vacuum.

In 1887, Heinrich Hertz succeeded in producing by electrical means waves that could be reflected, refracted, polarised etc. like ordinary light. Their wavelength was found to be much greater than that of longest infra-red rays (0.01 cm) described above.

Subsequently radio-waves of the very highest frequency the so-called micro-waves were produced with wavelengths as short as 1.8 mm and even still smaller, so that the gap between micro-radio-waves and infra-red waves no longer exists.

Below the ultra-violet region we come to *X*-rays produced in a discharge tube by the sudden stoppage of the electrons emitted from the cathode when they hit a target of hard metal. X-rays too were found to possess all the properties of light waves except that they were much shorter in wavelength of the order of 0.1 Å. It is this fact which accounts for the difficulty in reflecting and diffracting *X* rays. (For a detailed description of *X*-rays see some book on Modern Physics).

Similar to *X*-rays but more penetrating are the so called γ rays emitted by radioactive substances and in many sub-atomic processes. They are found to have wavelengths of 10^{-2} Å and less. They too have been shown to be of the same essential nature as light waves. Still shorter are the cosmic rays coming from the sun or interstellar spaces. These waves have great powers of penetration as their wavelength is extremely small of the order of 10^{-12} cm.

All these types of radiations described here have been shown to be waves of an electromagnetic character and we speak of the complete spectrum as electromagnetic spectrum. A complete chart of this spectrum and the part occupied by the visible light will be seen to be extremely small in comparison with the wavelengths known.

It may, however, be noted that while the various ranges are in general produced and detected in different ways and cause widely different effects, all consist of the waves of the same fundamental nature differing only in wavelength. The unification represented by the electromagnetic spectrum is one of the most impressive facts in physics and is the result of an enormous amount of research and investigation.

8

Interference of Polarised Waves

Elliptically and circularly polarised lights. Let a beam of plane-polarised monochromatic light of wavelength λ and amplitude A be incident normally upon a uniaxial doubly refracting crystal, with its faces cut parallel to the optic axis. Suppose the incident light is vibrating along PQ making an angle θ with the optic axis of the crystal. On entering the crystal the incident light will be split up into two components, ordinary and extraordinary. The ordinary component will have vibrations along PO and extra-ordinary along PE. The amplitudes of vibration of E and O rays arc $A\cos\theta$ and $A\sin\theta$ respectively along the optic axis of the crystal and perpendicular to it.

Both the components travel in the same direction, that is, perpendicular to the plane of paper, but with different speeds depending upon the nature of the crystal. In a negative crystal, such as calcite, E-ray travels faster than O-ray ($\mu_0 > \mu_e$), while in a positive crystal such as quartz, E-ray travels slower than O-ray ($\mu_e > \mu_0$) . It should be clearly borne in mind that, it is because of unequal velocities of the two components that although on entering the crystal their phases are same yet on emergence their phases are different as the two rays will have a path difference depending upon the thickness, and nature of the crystal and wavelength of the incident light. Thus the emergent components on leaving the crystal will consist of simple harmonic vibrations, in mutually perpendicular planes, having same period but different amplitudes and phases. They will be compounded into a single motion which can be linear, circular or elliptic depending upon the phase difference and the magnitude of θ.

Theory. If δ be the phase difference introduced between the two components in passing through the crystal, then by a convenient choice of the origin of time, their displacements in a negative crystal can be represented by

E-ray, $x = A \cos\theta \sin(\omega t + \delta)$

O-ray, $y = A \sin\theta \sin\omega t$

They can be put as

$$x = a \sin(\omega t + \delta)$$

and $y = b \sin \omega t,$

where $a = A \cos\theta$

and $b = A \sin\theta.$

On compounding these the path of the resultant vibration will be given by

$$\frac{x^2}{a^2} + \frac{y^2}{b^2} - \frac{2xy}{ab}\cos\delta = \sin^2\delta,$$

which represents the general equation of an ellipse with its major and minor axes inclined to the coordinate axes at certain angles depending upon their relative phases.

Case I. When $\delta = 0$, or $2n\pi$ the equation of the resultant motion becomes

$$\pm\left(\frac{x}{a}-\frac{y}{b}\right) = 0$$

or $$y = +\left(\frac{b}{a}\right)x,$$

which is the equation of a straight line passing through the origin and having a slope $\left(+\frac{b}{a}\right)$. This indicates that the resultant light is plane polarised in the *same* plane as the incident light.

Case II. (*i*) When $\delta = \frac{\pi}{2}$ or $(2n + 1)\frac{\pi}{2}$ and $a \neq b$ $[\theta \neq 45°]$ the equation of resultant motion reduces to

$$\frac{x^2}{a^2}+\frac{y^2}{b^2} = 1,$$

which is the equation of an ellipse. Thus the emergent light is elliptically polarised the plane of the ellipse being normal to the direction of propagation.

(*ii*) A special case arises when $\theta = 45°$, in addition to $\delta = \frac{\pi}{2}$ or $(2n + 1)\frac{\pi}{2}$, we then have $a = b$ and the equation of the resultant motion reduces to

$$\frac{x^2}{a^2}+\frac{y^2}{a^2} = 1,$$

or $$x^2 + y^2 = a^2,$$

which is the equation of a circle with radius a. Thus the emergent light is now circularly polarised.

Cage III. If $\delta = \pi$ or $(2n + 1)\pi$, the equation of the resultant motion reduces to

$$\pm\left(\frac{x}{a}+\frac{y}{b}\right) = 0$$

or $$y = -\left(\frac{b}{a}\right)x,$$

which is again the equation of a straight line through the origin with slope $\left(-\frac{b}{a}\right)$ in the negative direction of the X-axis. In this case the transmitted light is plane-polarised and its direction of vibration makes an angle equal to $\tan^{-1}\left(\frac{b}{a}\right)$ with that of the incident light.

For values of $\delta = \pi/4$ or $3\pi/4$, the equation represents the general equation of the ellipse but with major axis inclined in the positive and negative directions of the X-axis.

Indicate the shape of the resultant transmitted vibrations for different values of phase difference between the two rectangular components. For $0 < \delta < \pi$ the elliptical vibration is described counter-clockwise with respect to an observer towards whom the light travels and the light is said to be *left-handed* elliptically polarised light. For $\pi < \delta < 2\pi$, the ellipse is described in the clockwise direction and the light is said to be *right-handed* elliptically polarised.

Elliptic vibrations represent the general cases when the two components have any phase difference other than a multiple of π except for the special case of circular vibrations.

The results obtained under cases I to III above can be easily demonstrated by placing a thin wedge of quartz, cut with its optic axis at right angles to the principal section between two crossed nicols. It is assumed that the initial vibration make an angle of 45° with the edge of the wedge *For* monochromatic light the field of view will be crossed by dark and bright fringes. For $\delta = 0$, there will be darkness and for $\delta = (2n + 1)\pi$, there will be brightness. Between these two positions the light will be elliptically or circularly polarised.

Quarter-wave Plate

It is a uniaxial doubly refracting crystal plate, cut with its optic axis parallel to the refracting faces, and can produce a phase difference of $\pi/2$ or a path difference of $\lambda/4$ between the ordinary and the extra-ordinary rays.

When a beam of plane-polarised monochromatic light of wavelength λ is incident normally on a crystal of this type it is decomposed into O and E components which travel in the *same* direction but with *different* velocities.

In the case of *negative* crystal such as calcite E-ray travels faster than O-ray so that $\mu_0 > \mu_e$.

If t be the thickness of the plate then the distance travelled in the crystal is equivalent to distances μ_0, t and (μ_e t in air for the O and E rays respectively. Therefore the resultant path difference between the two components will be

$$= t(\mu_0 - \mu_e)$$

Since the phase changes by 2π for one wavelength λ, then to introduce a phase difference of $\pi/2$ between O and E rays is to introduce a path difference of $\lambda/4$, we have

$$\frac{\lambda}{4} = t(\mu_0 - \mu_e)$$

or $$t = \frac{\lambda}{4(\mu_0 - \mu_e)}$$

or $$t = \frac{\lambda}{4(\mu_0 - \mu_e)} \qquad \text{... (1)}$$

This is the thickness of a quarter-wave plate.

For a *positive* crystal such as quartz $\mu_e > \mu_0$,

so that $$t = \frac{\lambda}{4(\mu_e - \mu_0)}. \qquad \text{... (1a)}$$

A quarter-wave plate is the simplest device for producing and detecting circularly-polarised light. In conjuction with a nicol prism it is used for analysing all kinds of polarised light.

Half-wave Plate

A uniaxial doubly refracting crystal, cut with its faces parallel to its optic axis, which can introduce a phase difference of π or a path difference of $\lambda/2$ between the O and E rays is called a half-wave plate.

If t be the thickness of such a plate, then for a negative crystal such as calcite $\mu_0 > \mu_e$.

$$\therefore \qquad \frac{\lambda}{2} = t\,(\mu_0 - \mu_e)$$

$$t = \frac{\lambda}{2(\mu_0 - \mu_e)} \qquad \text{... (2)}$$

For a positive crystal such as quartz $\mu_e > \mu_0$.

$$\therefore \qquad t = \frac{\lambda}{2(\mu_e - \mu_0)} \qquad \text{... (2a)}$$

Since the path difference depends upon the wavelength of light used, the principal indices for yellow sodium light, $\lambda = 5893 \times 10^{-8}$cm. are generally used for computing the required thickness of a quarter-wave or a half-wave plate.

A half-wave plate finds its application in the construction of Laurent's half-shade device used in a polarimeter. Such plates are often made of thin sheets of quartz or of split mica cut parallel to the optic axis.

The following cases indicating the action of quarter and half-wave plates are worth noting.

(*i*) When plane-polarised light is incident on a quarter-wave plate, the emergent light is in general elliptically polarised; the axes of the ellipse are parallel and perpendicular to the optic axis and the ratio of the axes is given by tan θ, where θ is the angle which the plane of polarisation of the incident beam makes with the optic axis; When θ is 45°, the light emerging from a quarter-wave plate is circularly polarised.

(*ii*) When a plane-polarised light is incident upon a half-wave plate, the transmitted light is also plane-polarised. If the plane of polarisation of the incident light makes an angle θ with the direction of the optic axis, the plane of polarisation of the emergent beam makes an angle – θ with the same direction, *i.e.*, the plane has effectively been rotated through an angle 2θ.

Example. *Calculate the thickness of a quarter-wave plate for light of wavelength 5896 Å, the refractive indices for ordinary and extraordinary rays being 1.54 and 1.55 respectively.*

Let t be the required thickness of the plate, here $\mu_0 = 1.54$; $\mu_e = 1.55$; $\lambda = 5896 \times 10^{-8}$ cm.

Since $\mu_e > \mu_0$ the crystal is positive and the required thickness is given by

$$t = \frac{\lambda}{4(\mu_e - \mu_0)}$$

$$= \frac{5896 \times 10^{-8} \text{ cm}}{4(1.55 - 1.54)}$$

$$= \mathbf{0.001474 \text{ cm.}}$$

Example. *Calculate the thickness of doubly refracting-crystal required to introduce a path difference of λ/2 between, the O and E rays when λ = 8000 Å, $\mu_o = 1.65$ and $\mu_e = 1.48$.*

Here $\mu_0 = 1.65$; $\mu_e = 1.48$; $\lambda = 6000 \times 10^{-8}$ cm.

Since $\mu_0 > \mu_e$ the crystal is negative and the required thickness is given by

$$t = \frac{\lambda}{2(\mu_0 - \mu_e)}$$

$$= \frac{6000 \times 10^{-8} \text{ cm}}{2(1.65 - 1.48)}$$

$$= \mathbf{0.0001765 \text{ cm.}}$$

Production of elliptically polarised light. Before giving the experimental technique employed for the production of elliptically polarised light it is worthwhile to recall that *the elliptically polarised light results from compounding together two mutually perpendicular plane-polarised coherent light waves of unequal amplitudes but differing in phase by π/2, Case II (i).*

To obtain elliptically polarised light a beam of parallel monochromatic light, say from a sodium flame, is passed through two nicols which are arranged in the crossed position. For the sake of convenience the shorter diagonals of the end faces of the nicols are set mutually perpendicular to each other, that of the first, say, vertical and that of the other, horizontal. A quarter-wave plate is inserted between them. It is held normal to the incident beam in a suitable frame so that it can be rotated about a horizontal axis through any desired angle. Let its optic axis lie inclined at about 30° to the plane of vibration of the incident light.

The plate introduces a phase difference of $\pi/2$ between the two components *O* and *E* into which the plane-polarised light from the first nicol is split up. These components are of unequal amplitudes and on emergence from the quarter-wave plate give rise to an elliptical wave so that emergent light is elliptically polarised and in general on examination through the second nicol the field of view will be illuminated. On rotating the analysing nicol, the intensity of the emergent beam will vary between a *maximum* and a *minimum*.

Production of circularly polarised light. We have seen in

Case II (*ii*) that *circularly polarised light arises from compounding together two mutually perpendicular coherent vibrations of equal amplitudes and periods but differing in phase by* $\pi/2$. To get circularly polarised light the nicols are placed in the crossed position and the quarter-wave plate is rotated until the field of view is dark. This will happen when the optic axis of the quarter-wave plate is parallel to the "principal plane of the polarising nicol. Then the frame holding the plate is rotated through 45°. The quarter-wave plate is now in the desired position and the incident light is resolved into two equal components, the *E* and *0* vibrations respectively along and perpendicular to optic axis. Initially both the components are in the same phase but on emergence the *O* wave lags behind the *E* wave by one-quarter of a wavelength if the quarter-wave plate is of calcite. This is equivalent to a phase difference of $\pi/2$ between the *O* and *E* waves. On emergence from the quarter-wave plate they compound into a circular wave and hence the light is circularly polarised. If at this stage the analysing nicol is rotated the field of view shows no change in the intensity of the emergent beam; apparently the field is identical with that obtained when ordinary light passes through a nicol prism.

Polarised Lights

(a) *Plane-polarised light.* Allow the plane-polarised light to pass through a nicol prism capable of rotation about the incident beam as axis. When the direction of its vibration is parallel to the principal section of the nicol the beam will pass out and if it is perpendicular to the principal section of the nicol, the beam will be cut off by total internal reflection. Thus as the nicol is rotated there will be extinction of light, twice in each rotation of the analysing nicol in the case of plane-polarised light.

(b) *Circularly polarised light.* When a beam of circularly

polarised light is viewed through a nicol prism, then for *every* position of the nicol the circular vibration of the incident light is split up into two rectangular components of *equal* amplitudes. The direction of these vibrations will be parallel and perpendicular to the principal section of the prism. The parallel component will pass out while the perpendicular component will be totally reflected. As the amplitude of the transmitted beam is same for all positions of the nicol so that with rotation of the nicol there is no change in the intensity of the transmitted light. Thus the circularly polarised light gives the same effect as the unpolarised light and the two cannot be distinguished by a nicol prism alone.

Distinction between circularly polarised light and unpolarised light. Let a quarter-wave plate be interposed in the path of the beam of light. If it is circularly polarised it will be resolved into two rectangular vibrations of equal amplitudes, one parallel and other perpendicular to the optic axis of the quarter-wave plate and having a phase difference of $\pi/2$. This means that after passing through the quarter-wave plate a *further* phase difference of $\pi/2$ is introduced between the two O and E components comprising the circular vibration and the total phase difference between the two rectangular components therefore becomes $\pi/2 \pm \pi/2$, *i.e.*, π or 0 with respect to the original plane vibration which produced the circularly polarised light on passing through a quarter-wave plate suitably orientated.

With this new phase difference the two components produce plane-polarised light which will be refused transmission by a nicol when its principal plane is perpendicular to the direction of vibration. On the ordinary unpolarised light the quarter-wave plate has no such effect and when analysed by a nicol, it will not be extinguished.

(c) *Elliptically polarised light.* When a beam of elliptically polarised light is allowed to fall upon a nicol prism it is split up into two rectangular vibrations of unequal amplitudes with directions parallel to and perpendicular to the principal plane of the nicol prism. On rotating the nicol about the incident beam as axis the intensity of the transmitted beam changes from maximum to minimum according as the principal plane of the nicol is parallel to the major or the minor axis of the ellipse. This, however, is just what is observed *when a beam consisting of a mixture of plane polarised and ordinary light is examined by a rotating nicol, so to say, the elliptically polarised light resembles partially polarised light or a mixture of ordinary unpolarised light and plane polarised light.*

The vibrations consisting of a mixture of ordinary unpolarised light and plane polarised light can be represented. Since the vibrations of the ordinary unpolarised light are distributed in all possible directions perpendicular to the direction of propagation taken perpendicular to the plane of paper they can be represented by aa_1, bb_1, cc_1, etc. Also let AB represent the direction of vibration of the place polarised light.

When the principal plane of the analysing nicol parallel to AB the intensity of the transmitted beam is *maximum* being sum of that due to plane polarised component and *one-half* of that due to the ordinary light. When the principal plane of the analysing nicol is parallel to cc_1, the plane-polarised light component is totally cut off and the intensity of transmitted light is due only to one-half of that of the ordinary unpolarised light and so the intensity is *minimum*. Thus the intensity of this mixture changes from maximum to minimum according as the principal plane of the analysing nicol is parallel to or perpendicular to the plane of vibration of the plane-polarised light. The mixture thus behaves in the same manner as the elliptically polarised light when examined through a rotating nicol prism.

Distinction between elliptically polarised light and a mixture of unpolarised light and plane polarised light. To distinguish between the two types the light under examination is passed through a quarter-wave plate. If the incident light is elliptically polarised and the quarter-wave plate is placed with its axis parallel to the major or minor axis of the incident elliptic vibrations, it will be resolved into two rectangular vibrations of unequal amplitudes with a phase difference of $\pi/2$ between them. By passage through the quarter-wave plate a *further* phase difference of $\pi/2$ is introduced between its two rectangular component vibrations along its two axes so that the total phase difference between the two rectangular components becomes $\pi/2 \pm \pi/2$, *i.e.*, π or 0. With this new phase difference the two rectangular components on emerging from the quarter-wave plate combine together to give a linear vibration, *i.e.*, the transmitted light is plane-polarised. On being examined by a nicol prism its intensity is reduced to zero twice in one rotation of the nicol, when the principal section of the analyser is perpendicular to the plane of vibration of the emergent light.

On the other hand, if the incident light be a mixture of plane-polarised and ordinary unpolarised light then it is transmitted through a quarter-wave plate without modification and shows only variation in intensity from a maximum to a minimum and no extinction occurs on examination through a rotating nicol.

Fresnel's Rhomb

From the theoretical study of the phenomenon of total internal reflection of *plane-polarised* light from the boundary of a *dielectric* like glass, Fresnel was led to the conclusion that *at each total internal reflection, at a suitable angle, a glass plate introduces a phase change of n/4 between the two rectangular component vibrations into which the incident light may be resolved.* One of these components is in the plane of incidence and the other at right angles to it.

To verify his conclusion he constructed a parallelepiped of glass with angles, such that a ray of plane polarised light *AB* falling normally at one end after suffering total reflection internally at *B* and again at *C* at equal angles of incidence of 54°, emerges normally from the opposite end along *CD*. When the incident beam is polarised at an angle of 45° to the plane of incidence *ABC* it is resolved into two *equal* components, one parallel and the other perpendicular to the plane of incidence, each being treated separately. Total internal reflection at *B* introduces a phase change of $\pi/4$ between the two components, and the beam is converted into elliptically polarised light. Upon a second total reflection at C a further phase change of $\pi/4$ is introduced between them. Thus in all a phase change of

$$\frac{\pi}{4}+\frac{\pi}{4} = \frac{\pi}{2},$$

is introduced between the two *equal* components. The emergent beam should, therefore, be *circularly polarised.* This was found to be the case.

Conversely, if the incident light is circularly polarised, the rhomb produces a *further* change of phase of

$$\frac{\pi}{4}+\frac{\pi}{4} = \frac{\pi}{2},$$

between the two component vibrations at two internal reflections. Thus the total pnase difference between the two components amounts to

$$\frac{\pi}{2}+\frac{\pi}{2} = \pi$$

so that the emergent light is plane-polarised with its plane of vibration making an angle of 45° with the plane of incidence. Such a rhomb therefore enables us to distinguish between ordinary light and circularly polarised light.

By means of this rhomb we can also convert elliptically polarised light into plane-polarised, for if the axes of the elliptic vibrations respectively lie in and perpendicular to the plane of incidence *ABC*, the two internal total reflections will introduce a further phase difference of $\pi/2$ between the two component vibrations and emergent light should be plane-polarised. This light will be extinguished by a nicol prism and we can, therefore, distinguish between an elliptically polarised light and partially polarised light with the help of Fresnel's rhomb.

Thus, Fresnel's rhomb possesses all the properties of a quarter-wave plate and is preferable to it in working with white light, since the variation of index of refraction with different colours in glass is inconsiderable. There is, however, one disadvantage of the rhomb, that is, the emergent light changes its directions as the rhomb is rotated so it is difficult to follow it with other parts of the apparatus.

Babinet's Compensator

It is another useful device employed in the *production* and *analysis* of elliptically polarised light. Essentially this consists of two wedge-shaped quartz prisms *ABC* and *ADC*, placed with their hypotenuses in contact so as to form a small rectangular block *ABCD*. The prisms are cut so that the optic axes are parallel to the external faces *AB* and *CD* of the compensator but perpendicular to each other and also to the direction of propagation *PQ* of the transmitted light. One of the wedges is fixed and the other is arranged to slide with respect to it by means of a micro-meter screw. If *plane-polarised light falls normally on the face AB with the plane of vibration inclined at an angle* θ *to the optic axis, it will be broken up into E and O components parallel and perpendicular to AB respectively and travelling with different velocities.* If t be the thickness of the first prism, the path difference between the two components

$$= (\mu_e - \mu_0)\, t \text{ in the first prism.}$$

Now since the principal planes of two prisms are at light angles, the components will change *roles* in going from one prism to the other, that is, the ordinary ray in the first prism becomes an extraordinary ray in the second and *vice- versa.* The velocities are therefore inter-changed and if t' be the thickness of the second prism at the same point, the path difference

$$= -(\mu_e - \mu_0)\, t' \text{ in the second prism,}$$

so the *resultant path difference* obtained by adding the above two expressions is given by

$$\delta = (\mu_e - \mu_0)\,(t - t').$$

Along the centre P, where both paths are equal ($t = t'$), the effective path difference is zero and the emergent, light will be *plane polarised in the original plane.* On each side of P one, vibration will be behind or ahead of the other on account of having traversed different thicknesses in either prism. Thus the light emerging at Q can be made either *linearly, circularly* or *elliptically* polarised, that is, the compensator has the effect of a wave-plate of varying thickness. Thus it has the advantage over a quarter-wave plate in that the latter causes the path difference exactly for one particular wavelength whereas the compensator can be arranged to suit any wavelength as it enables us to produce phase difference of any arbitrary amount between two plane-polarised waves in mutually perpendicular planes.

Analysis of elliptically polarised light with the Babinet's compensator. An arrangement for calibeating the compensator and for studying the elliptically polarised light with its help with a view to finding its characteristics as regards *(i) phase: difference between its components (ii) position of the axes* and *(iii) the ratio of the axes.*

The compensator C is placed between crossed nicol N_1 and N_2, the polariser N_1 being so oriented that the plane-polarised light emerging from it is vibrating in a plane at 45° to the optic axes of the prisms. Owing to variation of the effective thickness along the pensator at distances from the mid-point for which the retardation is even multiple of $\lambda/2$, *i.e.*, λ, 2λ, 3λ, ..., the light is plane-polarised in the same plane as transmitted by N_1 and is, therefore, extinguished by N_2. Since the compensator is of constant thickness in the vertical direction, the field of view will be crossed by a series of equidistant parallel dark bands At positions midway between them, the path differences are odd multiples of $\lambda/2$, *i.e.*, $\lambda/2$, $3\lambda/2$, $5\lambda/2$,..., the transmitted light is also plane-polarised but vibrating in a plane making an angle 2θ with the incident plane of vibration where θ is the angle between that plane and the direction of optic axis in the first prism.

The nicol N_2 will not extinguish this light and if $\theta = 45°$, the light from these places will be completely transmitted by the nicol N_2 so that these regions will be very *bright*. Thus in general we have two systems of bands along which the light is plane-polarised and the planes of polarisation of the two are inclined at 2θ (compare action of a half-wave plate).

In between these positions the light will in general be elliptically polarised.

The type of polarised light transmitted by the compensator.

The places marked 1, 2, 3, indicate the positions of the dark bands for one set of fringes where the phase difference is a multiple of 2π.

With white light the bands are coloured like Young's interference fringes, the central one where the thickness of the two prisms is the same, being *dark*.

Calibration of the compensator. To calibrate the instrument the cross-wires carried by the system which causes relative displacement between the prisms are set on the central dark fringe and the white light is replaced by a monochromatic source of light. The movable prism is displaced parallel to itself with the micro-meter screw thus diminishing or increasing the difference of thickness $(t - t)$ under the cross-wires according to the direction of motion until one of the bands is on the cross-wires and the reading is noted. Then the screw is turned to bring the next band on the wires and the reading noted again. The difference (f) between these two readings corresponds to a change of 2π in phase. This step is repeated for a number of bands and mean f determined.

The instrument may now be used to determine the characteristics of *elliptically* polarised light.

(i) *Determination of the phase difference.* The source S and the nicol N_1 are now removed and the elliptical light to be examined is passed through the compensator and nicol N_2 or alternatively the plane-polarised light passing through N_1 is rendered elliptical by inserting a quarter-wave plate of the doubly refracting crystal. To find the phase difference between the two components of the elliptic vibration, first set the central dark band on the cross-wires using plane-polarised light. Then, when the incident light is elliptically polarised the central band will be displaced in the field of view. At this band the original phase difference ϕ between the two component vibrations has been neutralized by that introduced by the compensator. This phase difference is consequently found by adjusting the micrometer screw till the central dark band is restored to the cross-wires. If this displacement be x, then we have

$$\frac{\phi}{2\pi} = \frac{x}{f}$$

or $$\phi = \frac{2\pi x}{f}.$$

(ii) *Position of the axes.* We know that the phase difference of the component vibrations taken along the axes is 90°. Hence to find the direction of axes set the compensator so that the cross-wire is over the central dark band when the incident light is plane-polarised. Then turn the screw by an amount $f/4$, so that the line under the wire will correspond to a path retardation of $\lambda/4$ or a phase difference of $\pi/2$ between the two components of the emergent vibration. Now let the elliptically polarised light fall on the compensator. In general the central dark band will not be under the cross-wire. Rotate the compensator about the line of vision until the central dark band moves on to the cross-wire. It is often necessary to rotate the analyser also to obtain the maximum distinctness of the fringes. In this position the axes of the elliptic vibration are parallel to those of the quartz wedges.

(iii) *Ratio of the axes.* To find the ratio between the axes of the elliptic vibration where the elliptic vibration is shown as divided into two components OB and OC parallel to the axes of the compensator. In this position let the principal plane OA of the analyser make an angle θ with OB, the principal section of one of the prisms. The vibration after leaving the compensator must be in the direction BC perpendicular to OA, since it is extinguished by the analyser at the central band. Consequently, the tangent of angle θ which the principal plane of the analyser makes with the optic axis of one prism is equal to the ratio of the components OB and OC of the elliptic vibration. Then also $\tan \theta = a/b$, where a and b are the lengths of the axes of the ellipse parallel to OB and OC respectively. If $\theta = 45°$, the light is circularly polarised. It is easy to tell when the principal section of the analyser is parallel to *OB* or *OC* because then the interference bands disappear and the entire field is uniformly illuminated.

Interference of polarised light. We have already seen that when a beam of plane-polarised light is incident normally on a uniaxial doubly refracting crystal cut parallel to the optic axis, it is split up into two polarised components with vibration planes at right angles. Both these components do not suffer any separation into two distinct beams but as they travel with different velocities there occurs a phase difference on emergence depending upon the thickness of the plate. They compound into a linear or elliptical vibration according as the path difference is

$$n\pi \quad \text{or} \quad (2n + 1)\pi/2,$$

where n is an integer. As the beams are composed of simple harmonic motions at right angles, we do not expect interference phenomenon as with ordinary beams. If, however, the light after transmission through the crystal is made to pass through a nicol then a portion of each beam is transmitted with its vibration plane in the principal plane of the nicol and they would interfere like ordinary beams. The necessary conditions under which two beams of plane-polarised light may produce interference fringes were investigated by Fresnel and Arago in 1811 in a rather lengthy and complicated set of experiments which we need not describe here summarised as follows :

Their results may be

(1) Two beams of light polarised in the same plane produce interference in the same manner as ordinary light does.

(2) Two beams of tight, polarised in planes at right angles to each other do not produce interference fringes under any condition.

(3) Two beams of light polarised at right angles and then brought to the same plane of polarisation will interfere in the same manner, as ordinary light, only if they are coherent, that is, they originally belonged to the same beam of plane-polarised light or the same component of a beam, of unpolarised light.

(4) Two beams of plane-polarised light derived from-perpendicular component of unpolarised light and afterwards rotated into the same plane do not produce interference.

Colours due to thin crystalline plates. To observe the phenomenon of interference with polarised light let us place two nicols. N_1 and N_2, with their principal planes at right angles, *viz.*, in the *crossed* position so that no light is transmitted through them. Suppose now a thin plate *C* of a doubly refracting crystal is placed between them.

Let AB, represent the vibration plane of the plane-polarised light leaving N_1 with amplitude a and making an angle θ with the optic axis *XY* of the crystal plate. This amplitude a is broken up into an *E* component *HG* of amplitude $a \cos \theta$ along the optic axis *XY* with vibrations in the principal plane of the polarising nicol N_1 and with amplitude a and making an *O* component *HK* of amplitude $a \sin \theta$ with vibrations perpendicular to its principal plane.

Since they travel with different velocities, on emergence one will be ahead of the other in phase Here only the components *HL* of the *E* vibrations and *HM* of the *O* vibrations, both parallel to *CD*, will be transmitted where *CD* is the vibration plane of the vibrations transmitted through N_2. The resultant amplitudes of these components are

$$E' = E \sin \theta = a \cos \theta \sin \theta$$

$$O' = O \cos \theta = a \sin \theta \cos \theta.$$

The amplitudes are equal in magnitude when the nicols are crossed with a phase difference depending upon the thickness of the crystal plate. If the thickness is such that $\delta = 0, 2\pi, 4\pi$, the components interfere destructively so that with monochromatic light and a wedge-shaped crystal alternately dark and bright bands parallel to the thin edge of the wedge are seen. From this experiment we conclude that destructive interference is only produced when the two components are brought in the same plane after transmission through the second nicol in accordance with condition (3) stated above. If white light is used and a thin wedge-shaped plate of mica, calcite or quartz cut parallel to the axis is placed between crossed nicols, the interference pattern on the screen is brilliantly coloured.

This is due to the fact that the phase difference between *E* and *O* rays depends upon the wavelength as well as the thickness of the plate. If the phase difference at a point is 2π for, say, yellow light, it will not be so for other wavelengths ; the red and orange will suffer a phase Change less than 2π and green, blue, and violet more than 2π. Thus at this point yellow will be suppressed and the resultant colour will be a mixture of the remaining constituents.

Thus there will be a separate set of bands for each wavelength, the colour of the emergent light continually changing as we pass along the wedge. The colours will be brilliant when the nicols are crossed or parallel and will vanish when the principal section of the analyser is either parallel or perpendicular to the optic axis of the crystalline plate. It will also be observed that a rotation of the analyser through 90° will cause the colour Of each point to change to its *complementary*. This is obvious as the colour at any point is due to the suppression of certain wavelengths by the analyser, A rotation of the analyser through 90° allows the wavelengths previously transmitted to be suppressed and those previously suppressed to be transmitted.

Analytical treatment. The phenomenon of interference of plane-polarised light will now be discussed analytically and in a more general way.

Let OP and OA, represent the principal planes of the polariser and the analyser while OX represents the principal plane of the intervening thin crystal plate and OY a direction perpendicular to the latter. Let $\angle POX = \theta$, $\angle AOX = \phi$ and let $OK = a$, represent the amplitude of the light incident on the crystal plate. This is split up into an E beam of amplitude $a \cos\theta$ with vibrations parallel to OX and an O beam of amplitude $a \sin\theta$ with vibrations perpendicular to OX.

These two mutually perpendicular vibrations will travel with unequal velocities along the same direction so that on emergence from the crystal plate they have a phase difference δ given by

$$\delta = \frac{2\pi}{\lambda}(\mu_0 \sim \mu_e)\,t$$

where t is the thickness of the crystal and μ_0 and μ_e the refractive indices for ordinary and extra-ordinary vibrations respectively.

When these two components strike the analyser, they give rise to four sets of vibrations.

(*i*) $a \cos\theta \cos\phi$ vibrations parallel to OA, (*ii*) $a \cos\theta \sin\phi$ vibrations perpendicular to OA, (*iii*) $a \sin\theta \sin\phi$ vibrations parallel to OA and (*iv*) $a \sin\theta \cos\phi$ vibrations perpendicular to OA.

The components (*i*) and (*iii*) being parallel to OA are transmitted whereas components (*ii*) and (*iv*) are suppressed by the analyser.

The transmitted components (*i*) and (*iii*) being in the same plane and differing by a phase δ given above are compounded by the triangle law, [see equation (10)] the resultant amplitude being given by

$$A^2 = a^2\cos^2\theta\cos^2\phi + a^2\sin^2\theta\sin^2\phi + 2a^2\sin\theta\cos\theta\sin\phi\cos\phi\cos\delta$$

$$= a^2\Big[\cos^2\theta\cos^2\phi + \sin^2\theta\sin^2\phi + 2\sin\theta\cos\theta\sin\phi\cos\phi$$

$$- 4\sin\theta\cos\theta\sin\phi\cos\phi\sin^2\frac{\delta}{2}\Big]$$

$$= a^2\left[(\cos\theta\cos\phi + \sin\theta\sin\phi)^2 - 4\sin\theta\cos\theta\sin\phi\cos\phi\sin^2\frac{\delta}{2}\right]$$

$$= a^2\left[\cos^2(\theta-\phi) - \sin 2\theta\sin 2\phi\sin^2\frac{\delta}{2}\right] \quad \ldots (3)$$

Since the intensity is proportional to the square of the amplitude, this expression enables us to study the variation in intensity due to changes in δ, θ and ϕ.

It consists of two terms. The first term is independent of 8 and hence of the wavelength giving a background illumination but in-the second term δ will depend upon wavelength and consequently different colours, etc. will vary in strength. With monochromatic light alternate dark and bright bands will be produced. With white light, wavelengths for which $\delta = 2n\pi$, *viz.*, 0, 2π, 4π, etc. will be suppressed while wavelengths for which $\delta = (2n + 1)\pi$, *viz.*, π, 3π, 5π, etc. will be present, the crystal will appear to be uniformly coloured except when θ and ϕ are either zero or $\pi/2$ simultaneously or separately.

The following cases are of *special* interest:

(1) When the polariser and the analyser ate in fixed orientasions, that is, θ and ϕ are constant, then the intensity varies with the value of $\sin^2\delta/2$, δ depending upon the thickness of the crystal plate and the wavelength of light. The results deduced earlier follow at once.

(2) When the nicols are crossed, *i.e.*, $\theta - \phi = 90°$, Then the intensity

$$I_a = kA^2 = ka^2\left[0 - \sin 2\theta \sin\left(2\theta - 180°\right)\sin^2\frac{\delta}{2}\right]$$

$$= ka^2 \sin^2 2\theta \sin^2\frac{\delta}{2}. \quad \text{... (4)}$$

Here the background illumination is absent and we have the same colour effects as in case (1) above. The colours in this case are real and pure. For a given value of θ, the intensity I is maximum when $\delta = (2n + 1)\pi$ and zero when $\delta = 2n\pi$.

(3) When the nicols are parallel, *i.e.*, $\theta = \phi$, then the intensity

$$I_p = kA^2 = ka^2\left[1 - \sin^2 2\theta \sin^2\frac{\delta}{2}\right] \quad \text{... (3)}$$

In this case the first term in the expression has maximum value indicating that the background illumination is well marked. The colours are complementary to those in case (2) since $I_p + I_0 = ka^2$.

Interference with highly convergent light — Rings and brushes. An arrangement for studying the interference pattern with highly convergent light. A plate C of a uniaxial crystal of uniform thickness is placed in the path of a plane polarised beam of light rendered convergent by a short focus lens L_1. After passing through the crystal, the light is made parallel again by a similar lens L_2.

A lens L' placed beyond N_2 focuses on the screen X all the rays leaving C. In thus case a cone of light 1-2 is incident on the plate and the different rays of the cone are differently inclined to it so that the O and E components undergo different changes in phase in traversing it. If now the crystal is cut with the optic axis perpendicular to the face on which the light is incident the O and E components travel with equal velocities along the axis of the beam of light through O and no phase change takes place. We find that the rays 1 and 2, travel a greater distance in the crystal and being inclined to the optic axis are doubly refracted.

Since their velocities are different there will be a phase difference between *O* and *E* rays which depends upon the angles of incidence and refraction. Consider a point *P*, on the face of the crystal supposed parallel to the plane of the paper. All rays entering at points on the circumference of circle of radius *OP* pass through the same crystal thickness and show the same phase retardation on emergence. Let *OY* and *OX* be drawn parallel to The principal plane of the polariser and the analyser respectively.

The incident beam of amplitude *PQ* is broken up into an *E* component with vibrations in the principal plane and, therefore, parallel to *PE* and an *O* component with vibrations perpendicular to *PE*, *viz.*, along *PO*. On arriving at the second nicol N_2 the components *PE* and *PO* will be transmitted and interfere destructively or otherwise depending upon the phase difference on emergence. The magnitude of the phase difference for a crystal of uniform thickness will be the same for all points on a circle with centre *O* and radius *OP*. For points on another circle, the phase will be different. If the plate is a few millimetres thick, there will be a series of regularly spaced concentric circles where the path difference will be some multiple of 2π to cause destructive interference.

So the transmitted light will consist of *concentric rings of interference*. In addition to this since the principal section of the analyser is at right angles to *OY* (direction of vibration of incident light) the light emerging from points on the straight lines *OX* and *OY* will be stopped and the field of view will be crossed by two *black brushes* at right angles to each other. If the second nicol is parallel to the first the pattern is reversed in every respect. The dark brushes and rings of the first pattern are bright and the bright brushes and rings dark. This means that part of the light which was intercepted by the second nicol

N_2 in the crossed position is transmitted for the parallel position. Similarly, the light which passed through in the first case is (now intercepted).

If incident light is *white,* the bright rings corresponding to the shorter wavelengths will possess smaller diameters than those corresponding to longer wavelengths and hence there will be a series of *coloured rings* in the field of view concentric with O and on these rings, the crosses or brushes black or white according as the polariser and analyser are crossed or parallel will be superposed.

9

Nature of Light Waves

Nature of light waves. Experiments in the preceding chapters on interference and diffraction have shown beyond any shadow of doubt that light is some form of wave motion. But in explaining these phenomena on the wave theory we made no tacit assumptions regarding the nature of light vibrations. Naturally the question arises, are the light waves *longitudinal* or *transverse*? We shall, therefore, investigate the nature of these waves and to begin with, discuss an important feature which distinguishes the two types. From the study of sound we know that a longitudinal wave has the same properties with respect to any plane through its line of propagation while a transverse wave behaves differently in different planes.

This statement can be illustrated by a simple mechanical analogy. Take a stretched rubber cord AB passing through two narrow slits S_1 and S_2 parallel to each other in the vertical plane. Set up a longitudinal wave by moving the end A to and fro parallel to the length of the cord. Rotate any of the slits about AB as axis. It will be seen that this rotation does not affect the passage of the wave. On the other hand, if we set up a transverse wave in the cord by moving the end *A* up and down in the vertical plane, it will be seen that the wave passes through the slits undisturbed. Now rotate the slit S_2 through 90° about *AB* as axis, it will be noticed that the wave passes through the slit S_1 undisturbed as before but it is not able to pass through S_2 *so* that beyond S_2 the cord remains practically undisturbed. This shows that the transverse wave can pass through the slits only when they are parallel to each other and lie in the same plane in which the wave lies. Thus transverse waves have different properties with respect to the vertical plane as compared with other planes. Such waves are said to be polarised after passing through the first slit. This analogy helps us in distinguishing between the longitudinal and transverse waves.

The tourmaline and several other transparent crystalline substances behave with respect to light waves in exactly the same manner as the slits do with respect to transverse waves in the cord.

Take a tourmaline crystal *A* cut parallel to its crystallographic axis which is shown by the vertical doted line. Note that the crystallographic axis of a crystal is not a fixed line in the crystal but merely a direction and any line parallel to it is an identically equivalent axis in all respects. It is such a direction in crystals to which their shape and other properties may be referred in the simplest possible manner.

Hold the crystal normally to the path of a beam of light from a source *S*. The emergement beam will be slightly coloured. Rotate the crystal *A* about *SO* as axis it will be seen that there is no change in the intensity or the colour of the emergent beam. Take a second crystal *B* cut similarly and hold it in the path of the beam so that its axis is *parallel* to the axis of the crystal *A*. The beam of light will be seen to pass through both the crystals and will be still coloured. Now rotate the crystal

B about the beam as axis. It will be seen that the intensity of the emergent beam decreases and when the axes of both the crystals are at right angles to each other no light comes out of the crystal *B*. If the crystal *B* is further rotated light reappears and its intensity becomes maximum again when their axes are parallel. This occurs after a rotation of *B* through 90° from its position. Comparing the observations in this experiment with its mechanical analogue discussed earlier we can firmly conclude that the light waves are transverse, that is, *the vibrations in them are at right angles to the direction in which the wave is travelling.*

Plane polarised light. In case of light coming out of the tourmaline crystal *A*, the vibrations have acquired two-sidedness and are confined to a single plane perpendicular to direction of propagation of light. This light which acquires the property of two sidedness is 'called linearly polarised light'. Since the vibrations constituting the beam of light are confined only to one definite plane through the axis of the beam, the light is generally said to be 'plane polarised'.

The properties of a plane polarised beam differ with respect to two planes, one containing the vibrations and the other at right angles to it. The plane ABCD, containing the crystallographic axis in which the vibrations are occurring is known as the 'plane of vibration'. The plane EFGH at right angles to the plane of vibration of the light waves is called the 'plane of polarisation' vibration planes.

Representation of light-vibrations. A light wave is a transverse wave and has its vibrations at right angles to the direction of propagation. An ordinary beam of light, as it consists of millions of such waves each with its own plane of vibration, will contain waves vibrating in all possible planes with equal probability. The direction of propagation of a beam of light is perpendicular to the plane of the paper while its vibrations are in the plane of the paper.

A beam of ordinary light may be regarded as consisting of two sets of waves, one set vibrating in one plane and the other at right angles to it. The direction of propagation of the beam is in the plane and these two sets of waves vibrating at right angles to each other, one in the plane of paper and other perpendicular to it, are respectively represented by arrows and dots. The vibrations in the plane of the paper represented by arrows are polarised in a plane perpendicular to the plane of paper while those in the plane perpendicular to the plane of the paper and represented by dots are polarised in the plane of the paper.

Since in the case of ordinary light the vibrations in ether are supposed to be distributed over all possible planes containing the ray, so that in polarised light if the vibrations are *linear* the light is said to be plane polarised as explained in the preceding section. If the vibrations are *circular,* light is called circularly polarised and if they are *elliptic* the light is elliptically polarised. It is worth noting as shown already in the chapter on Wave Motion, that *circular and elliptic vibrations arc the resultant of two linear vibrations perpendicular to each other differing in phase by* $\pi/2$. The converse is also true.

Polarisation by Reflection

The simplest method of obtaining plane polarised light is by *reflection.* This method was discovered by Etiennie Louis Malus of France in 1808. After long experimentation he found that when a beam of ordinary light is reflected from the surface of a transparent medium like glass or water the reflected light is *partially* polarised and *the degree of polarisation varies with the angle of incidence.* The percentage of the polarised light in the reflected beam is greatest when it is incident at an angle known as the angle of polarisation or polarising angle for the medium. This angle of maximum polarisation which for ordinary glass is 57.5° varies slightly with the wavelength of the incident light, since refractive index depends upon the wavelength.

Thus complete polarisation is possible only with monochromatic light and not with white light. *The vibrations of this plane polarised reflected light are found to be perpendicular to the plane of incidence and, therefore, the reflected light is ' said to be plane-polarised in the plane of incidence.*

The above conclusions can be verified by means of the arrangement. Let a beam of ordinary light be incident along *PQ* on the reflecting surface *XY* of a transparent medium so that it is reflected vertically upwards along *QR*. Hold a tourmaline crystal *T* in the path of the reflected beam and rotate it about *QR* as axis. The examination of the transmitted light shows variation in intensity, indicating thereby that the reflected light *is partially* polarised. Keep the tourmaline crystal fixed in the minimum intensity position and *vary* the angle of incidence by changing the direction of the incident beam *PQ* and the inclination of *XY* until the intensity of the reflected beam after transmission through the crystal is *almost zero.* At this stage the percentage of the plane-polarised light in the reflected beam is maximum. The angle of incidence at which this occurs is found of be 57.5° in the case of glass surface and is called the polarising angle for the glass-air medium. It may be noticed that if now the tourmaline crystal is rotated about *QR* as the axis, the intensity of the reflected beam after transmission through the crystal varies between a *maximum* and a *minimum* twice in each rotation It is maximum or minimum according as the crystallographic axis of the tourmaline is perpendicular to or lies in the plane or incidence. This shows that the light vibrations in the plane polarised reflected beam are *perpendicular* to the plane of incidence. This gives us a simple method of production and detection of the plane-polarised light.

The detection of the plane-polarised reflected light can also be done by using another mirror instead of a tourmaline crystal. This is achieved in *Biot's polariscope.*

Biot's Polariscope

The principle of Biot's polariscope. Light from a monochromatic source S is allowed to fall at the polarising angle of incidence, upon a transparent glass plate A which is capable of rotation about a horizontal axis. By suitable adjustment of A, the light is reflected vertically upwards to a *black* glass plate B which can rotate both about the horizontal and vertical axes. The angle of rotation of B about the vertical axis can be noted on a horizontal scale attached to the frame holding it. The plate B is made from black glass to avoid internal reflections.

Both the plates A and B are held equally inclined to the horizontal so that light always falls upon them at the same angle of incidence. This means that the planes of incidence of the two reflecting surfaces A and B are parallel to each other. In this position the reflected beam is sent along BC and its intensity is maximum.

When the plate B is rotated about the vertical axis the intensity of the reflected beam goes on decreasing until when the planes of incidence of A and B are at right angles to each other. The intensity of the reflected beam is minimum, and for the polarising angle of incidence the intensity in this position of B is zero.

It is evident that in the first case the vibrations of the ray ABC are perpendicular to the plane of incidence of both A and B, while in the second case the vibrations of the ray AB are perpendicular to the plane of incidence of A and parallel to the plane of incidence of B, so that the light is not reflected from B. Thus the plate A polarises the incident light and the plate B analyses it. The explanation for polarisation is given in the following para.

The vibrations of the ordinary incident light which occur in all possible planes can be supposed to be resolved into two mutually perpendicular directions, one set being in the plane of incidence and the other normal to it. The vibrations in the plane of incidence shown by double arrowed lines perpendicular to the incident ray *SA* make different angles with the reflecting surface as the angle of incidence is varied. Thus when the angle of incidence equals the polarising angle for the media, they are *transmitted in toto* without giving rise to the reflected ray. While the other set of vibrations shown by dots always remains parallel to the reflecting surface whatever the angle of incidence and so it will be reflected for all angles of incidence. This reflection is, however, partial as some of these components also pass through. Thus the reflected light is *completely* plane polarised while the transmitted light is not so, as it also contains some of the vibrations perpendicular to the place of incidence.

Brewster's Law

Sir David Brewster, a Scottish physicist (1781-1868) as a result of series of experiments made with different reflectors discovered that there was a simple relation between the angle of maximum polarisation and the refractive index of the medium. This relation called Brewster's law is given by

$$\mu = \tan i,$$

where i is the polarising angle and μ the index of refraction. A direct deduction from this law is that *when light is incident at the angle of maximum polarisation the reflected ray is at right angles to the refracted ray.*

Let *XY*, be the surface of separation of a transparent medium, say glass, *PQ* the beam of light incident at the polarising angle, *QR* the reflected beam and *QS* the refracted beam.

Let i and r be the angles of incidence and refraction respectively.

From Brewster's law

$$\mu = \tan i = \frac{\sin i}{\cos i}.$$

From Snell's law

$$\mu = \frac{\sin i}{\sin r},$$

$$\because \quad \frac{\sin i}{\cos i} = \frac{\sin i}{\sin r}$$

or $$\frac{\sin i}{\sin (90 - i)} = \frac{\sin i}{\sin r}$$

$$\therefore \quad 90° - i = r$$

or $$i + r = 90°,$$

that is, at polarising incidence the reflected and the refracted rays are at right angles.

Brewster's law is obeyed even when light is reflected at the surface of a rarer medium. Thus if light is incident on a glass plate at the polarising angle, *i.e.*, 57.5° the refracted ray will be incident at the polarising angle on the lower surface of the plate. Obviously, the angles of incidence and refraction at the lower surface are r and i respectively. Then the refractive index of the second medium with respect to the first is $\frac{\sin i}{\sin r} = \frac{1}{\mu}$ and also

$$\tan r = \frac{\sin r}{\cos r} = \frac{\sin r}{\sin i} = \frac{1}{\mu}$$

since i is the angle of polarisation at the first surface and, therefore $\cos r = \sin i$.

Thus the light reflected from both the upper and lower surfaces of a glass plate will be polarised in the plane of incidence. This gives us a very convenient method for obtaining plane-polarised light. But this method is not very efficient because of a greater part of light being transmitted so that a number of glass plates placed one above the other can be used with advantage. This arrangement is called a pile of plates and is generally used for getting plane-polarised light.

The dots and arrows on the ray PQ indicate that the random vibrations in the incident beam are confined to two planes normal to the ray. The dots show the vibrations normal to the plane of incidence. Let us call it group (*a*) vibrations. The arrows show the vibrations normal to the ray but in the plane of incidence, say, group (*b*) vibrations. The experiments indicate that the reflected beam contains vibrations belonging to group (*a*), while the refracted beam contains those belonging to group (*b*) together with a certain proportion of the vibrations of group (*a*). The case is understandable if we remember that vibrations of group (*a*) being parallel to the refracting surface, *whatever* the angle of incidence, will be reflected for all angles of incidence while those of group (*b*) will make various angles with the surface as the angle of incidence is varied and will be transmitted without giving rise to a reflected ray when incident at a *particular* angle called polarising angle.

Example. *A ray of light is incident on the surface of a glass plate of refractive index 1.5 at the polarising angle. What is the angle of refraction ?*

Suppose i is the polarising angle of incidence and r the corresponding angle of refraction.

From Brewster's law

$$m = \tan i$$

$$\therefore \quad i = \tan^{-1} (\mu) = \tan^{-1} (1.5) = 56.3°.$$

$$\text{Since} \quad i + r = 90°$$

$$\therefore \quad r = 90° - i$$

$$= 90° - 56.3° = 33.7°.$$

Polarisation by a pile of plates. According to Brewster's law when a beam of light is incident on a thin glass plate at the polarising angle, the light reflected from both the upper and lower surfaces of the plate is *completely* plane-polarised in the plane of incidence, *i.e.*, having vibrations perpendicular to the plane of incidence. The intensity of this reflected beam is very weak as 100% of the light vibrations parallel to the plane of incidence and 85% of the perpendicular vibrations are transmitted and only 15% of the perpendicular vibrations are reflected, *i.e.*, about 15/185 = 8 % of the incident light is reflected at each reflection. Obviously, therefore, to increase the intensity of plane polarised reflected beam we should increase the number of reflections and consequently the number of plates. Accordingly a set of about ten thin plates is laid one upon another and arranged in a wooden tube with their planes at an angle of 33° to its axis so that if a beam of light parallel to the axis of the tube falls on the pile, it will be incident at an angle of 57°, the polarising angle for glass.

By this arrangement earlier designated as pile of plates, more and more of the vibrations perpendicular to the plane of incidence are reflected and all those parallel to it are transmitted as a refracted beam. Thus the reflected plane-polarised beam increases in intensity with the increase in the number of plates, while the transmitted beam loses more and more of its perpendicular vibrations, consequently the percentage of the plane-polarised light in the transmitted beam also increases. The larger the number of surfaces, the greater is the intensity of the reflected plane-polarised beam and also more nearly is the transmitted light plane-polarised. This also indicates that the transmitted light is *never* completely plane-polarised for any angle of incidence, because it would always contain some vibrations in the plane perpendicular to the plane of incidence.

A similar pile can be employed as an analyser, that is, for

detecting plane-polarised light. When the planes of incidence of the two piles, the polariser and the analyser are parallel, the intensity of the reflected beam is maximum, whereas when their planes of incidence are perpendicular, its intensity is minimum.

Doable refraction. When a beam of ordinary un-polarised light is allowed to pass through a calcite or quartz crystal there are two refracted beams in place of the usual one as in glass. The crystals having this property are said to be *doubly refracting* or to exhibit *birefringence* and the phenomenon is called double refraction. It was first of all discovered by Erasmus Bartholinus, a native of Denmark, in 1669.

To illustrate this phenomenon allow a narrow beam of light from a point source to pass through a crystal of Iceland spar or calcite, two images will be obtained on the screen.

The phenomenon can also be illustrated in a simple manner by putting a dot on a sheet of paper and viewing it through a calcite crystal. Two images will be seen. *On* rotating the crystal about the incident beam as axis one image remains stationary while the other rotates around the first. The stationary image is known as the *ordinary* image and the refracted rays which produce this image are known as ordinary rays because they obey the ordinary laws of refraction. The other image is called *extra-ordinary* image and the refracted rays which give rise to this image are called extra-ordinary rays, because they do not obey the ordinary laws of refraction. Since the two opposite faces of a calcite crystal are always parallel so that the two refracted rays, ordinary and extra-ordinary, emerge parallel to the incident beam and, therefore, parallel to each other. For normal incidence the ordinary ray will pass straight through the Crystal without deviation whereas the extra-ordinary ray will be refracted at some angle and will come out paraliel to and displaced from the incident ray. As indicated above on rotating the crystal about the beam as axis, the *E*-ray rotates about the fixed *O*-ray, the *line joining the O and E images always remaining parallel to the shorter diagonal of the emergent crystal face.*

It may be noticed that both the refracted rays—ordinary and extraordinary—are plane-polarised, and their planes of polarisation are at right angles to one another, the vibrations of ordinary ray being normal to the plane of the paper while those of the extraordinary take place in the plane of the paper. This can be verified by viewing the images through a tourmaline plate. On rotating the tourmaline plate the intensities of both the images undergo a change. If the intensity of the ordinary image *if* decreasing that of extraordinary image is increasing and *vice versa,* so to say if one is at its brightest the other is not visible. In short, the ordinary and the extraordinary images alternately are at their brightest twice in one complete rotation.

Calcite. Calcite also known as Iceland spar, is transparent to visible as well as to ultraviolet light. It is chemically calcium carbonate ($CaCO_3$) and occurs in nature in different forms, all of which give rhombohedron on cleavage. Each of the six faces of this crystal is a parallelogram having angles of 78° and 102° nearly. At the two diametrically opposite corners *A* and *G* the angles of the three faces meeting there are all obtuse while at its remaining six corners one angle is obtuse and two are acute. The corners *A* and *G* are known as *blunt* corners.

Optic Axis

An important feature concerning the behaviour of calcite and quartz is that there is a direction through these crystals along which *there is no double refraction.* This particular direction is called the optic axis.

In calcite if a line is drawn through one of the corners (*A* or *G*), making equal angles with the three edges meeting there it would represent the direction of the optic axis and any other Fine parallel to it is an identically equivalent axis in all respects.

If the rhombohedron is so cut that all its edges are equal then *the line joining the opposite blunt corners (AG) or any line parallel to it gives the direction of the optic axis.* It may be clearly noted that the optic axis is a direction and not a particular line. A ray passing along the optic axis does not break up into ordinary (O) and extraordinary (E) rays, i.e., their velocities remain, the same along the optic axis .More so, the crystal is symmetrical about the optic axis.

Principal section and principal plane. A plane containing the optic axis of the crystal and perpendicular to the two opposite refracting faces is called a principal plane or principal section of the crystal. It may be noted that in this case the line joining the two images *O* and *E* is *always* parallel to the principal section of the crystal. For every point there are three principal sections, one for each pair of opposite crystal faces. A principal section always cuts the surfaces of a calcite crystal in a parallelogram having angles of 71° and 109°.

The principal plane of the ordinary ray is defined as a plane in the crystal which contains the optic axis and the ordinary ray and the principal plane of the extraordinary ray is a plane in the crystal which passes through the optic axis and the extraordinary ray. The ordinary ray always lies in the plane of incidence while this is not generally true for the extraordinary ray. The principal planes of the two refracted rays do not coincide except in special cases. The special cases are those for which the plane of incidence is a principal section.

Note. The ordinary ray is plane-polarised in the principal plane of the crystal while the extra-ordinary ray is plane-polarised in a plane at right angles to it, *i.e.*, the ordinary ray vibrations are perpendicular to the principal plane of the crystal while the extraordinary ray vibrations are in its principal plane.

Types of crystals. A class of crystals, such as calcite, quartz, ice, tourmaline and nitrate of soda in which there is a single direction known as the optic axis along which all waves are transmitted with one uniform velocity while in any other direction there are two velocities, are called uni-axial crystals. Another class of crystals such as borax, mica, topaz, aragonite and selenite in which there are two directions of one uniform velocity (*i.e.*, they have two optic axes), are known as bi-axial crystals. This latter class was discovered by Brewster. We shall confine ourselves only to the former class of crystals which is a special case of the latter class.

Polarisation by double refraction. The phenomenon of polarisation by double refraction was discovered in calcite by Huygens in 1678. In this case both the beams are plane-polarised in perpendicular planes as said above. If instead of one, two crystals are placed in the path of the beam with their principal planes parallel to each other, the two resultant beams arc separated by a distance equal to the sum of the displacements produced by each crystal, when used separately. Upon rotating the second crystal about the incident beam as axis each of the two beams *O* and *E* is split into two parts giving rise to four emergent beams.

At 90° rotation we again have two beams but the images change roles, the ordinary of the first becomes the extra-ordinary of the second and *vice versa*. Further rotation finds the originals reappearing and eventually if the crystals are of equal thickness these come together into one single beam in the centre for the 180° position.

Explanation. Ordinary light on entering the first crystal is broken up into two plane-polarised beams, one *O*-ray vibrating perpendicular to the principal plane and the other *E*-ray vibrating in the principal plane. So to say, the crystal resolves the incident light into two components, causing one type of vibrations to travel one path and the other to travel another path. Now consider one of the rays, say *E*-ray, to be incident on the second crystal making an angle θ with its principal plane. Let a be the amplitude of this incident ray. This second crystal like the first transmits light vibrating in its principal section along one path and that vibrating at right angles along another path. The ray is, therefore, split up into two components E_0 and E, with amplitudes $a \sin \theta$ and $a \cos \theta$ and emerges out with intensities $a^2 \sin^2\theta$ and $a^2 \cos^2 \theta$ respectively. At $\theta = 90°$, E_0 disappears and E attains maximum intensity of a^2. In short at all positions the sum of the components, $a^2 \sin^2 \theta + a^2 \cos^2 \theta$, is just equal to a^2, the intensity of the incident beam. The same treatment holds for the ordinary beam.

Thus plane-polarised light may be gainfully obtained by double refraction, and the most important method of producing and studying plane-polarised light by double refraction is by the use of crystals of calcite and quartz.

Law of Malus

It was experimentally shown by Etiennie Louis Malus in 1809 that *the intensity of the incident polarised ray is equal to the sum of the intensities of the two refracted rays.* This is known as law of Malus and can be simply accounted for by the principle of resolution of vectors. Let the vibration of the incident plane polarised beam having an amplitude a be along AP, making an angle θ with the principal plane of the crystal. The amplitude of E-ray is

$$x = a \cos \theta.$$

The amplitude of the O-ray is

$$y = a \sin \theta.$$

Since intensity is proportional to the square of the amplitude,

$$\therefore \quad I_0 = a^2 \sin^2 \theta; \quad I_e = a^2 \cos^2 \theta,$$

whence $$I_0 + I_e = a^2 (\sin^2 \theta + \cos^2\theta) = a^2 = I, \quad \text{... (2)}$$

where I is the intensity of incident polarised beam.

Thus the sum of the intensities of the transmitted rays is equal to the intensity of the incident ray.

The same treatment holds for both *O* and *E* rays which after passing through a second calcite crystal are further split up into their four plane-polarised components (*O* and *E* components of each).

To sum up, if a pencil of light of intensity *I* falls upon a crystal of Iceland spar it is divided into ordinary and extra-ordinary rays, each of average Intensity ½*I* and if these fall upon a second crystal, they will again be sub-divided according to the law of Malus as indicated in the following table :

Incident ray	*Transmitted by first crystal*	*Transmitted by second crystal*	*Place of polarisation*	*Sum*
	$O = \frac{1}{2} I$	$O_0 = \frac{1}{2}I \cos^2\theta$	0	
		$O_e = \frac{1}{2}I \sin^2\theta$	90	½*I*
		$E_0 = \frac{1}{2}I \sin^2\theta$	0	*I*
	$E = \frac{1}{2}I$	$E_e = \frac{1}{2}I \cos^2\theta$	90	½*I*

Nicol Prism

It is an ingenious optical device made from a calcite crystal for producing and detecting plane polarised light ; in the former case it is called a *polariser* and in the latter an *analyser*. It derives its name from its inventor William Nicol who first designed it in 1828. It is used in many optical instruments.

We have seen above that when a beam of ordinary un-polarised light is passed through a doubly refracting crystal, it is split up into two plane polarised beams —ordinary and extra-ordinary, whose planes of polarisation are mutually perpendicular. It is only necessary to get rid of one of them to obtain a strong beam which shall be plane-polarised.

In the Nicol prism the *O*-ray is eliminated by total internal reflection so that we are left with only *E-ray, whose vibrations lie in the principal plane.*

One of the most common forms of the Nicol prism is made by taking a long, clear rhomb of calcite so that its length *AE* is three times its width *AD*. The end faces *ABCD* and *EFGH* of the rhomb are ground so as to reduce the angles *C* and *E* from 71° to 68°. The resulting crystal is then cut into two parts along the plant A_1G_1, passing through the blunt corners and perpendicular to both the principal section and the end faces, so that A_1G_1 makes an angle of 90° with ends G_1E and A_1C. The two cut faces are ground and polished optically flat. They are then cemented by a thin layer of canada balsam which is a clear transparent substance whose index of refraction lies mid may between the indices of *O*-ray and *E*-ray of calcite. For example, for sodium light, λ = 5893 Å.

$$\mu_0 = 1.65837, \ \mu_{CB} = 1.550, \ \mu_0 = 1.48641.$$

It may be noted that the principal section of the crystal coincides with the principal planes of both the *O* and *E*-rays. The two end faces of the prism are kept open while its sides are coated with lamp black and kept covered by a brass tube. A ray incident along *SP* on one of the end faces of the nicol and moving nearly parallel to its length is divided into *O* and *E* rays whose vibrations are respectively perpendicular and parallel to the principal section of the nicol prism. The ordinary ray is travelling from a denser to an optically rarer medium and will suffer total internal reflection provided the angle of incidence at the balsam layer is greater than the *critical angle* for the two media, calcite and balsam,

$$i.e., \text{ greater than } \sin^{-1}\left(\frac{1.550}{1.658}\right) = 69°.$$

This totally reflected ray is absorbed by the lamp black layer on the side of the prism. The extra-ordinary ray on the other hand, travels from an optically rarer to an optically denser medium and so emerges out of the crystal, and leaves the faces EG_1 parallel to the direction SP of the incident ray. Thus we obtain a beam of plane polarised light of considerable intensity. Since the vibrations in the extra-ordinary ray occur in the principal plane, it follows that the vibrations in the emergent beam will be parallel to the plane of the paper. The end view of a nicol's prism. The direction of vibration in the emergent plane-polarised light is parallel to the shorter diagonal AC of the end face or in the direction of double arrow.

It may be noted that the angle of incidence on the face A_1C is limited to a very narrow range of about 14°. For the angle of incidence greater than this, the O-ray will strike the balsam layer at angles less than the critical and so will be transmitted.

Since only along the optic axis the E-ray travels with the same velocity as the O-ray and in all other directions it has different velocities, it being maximum at right angles to the optic axis ; the value of μ_e for sodium light is 1-4-6 or 1.658 according as the E-ray travels at right angles to the optic axis or along the optic axis. Along the optic axis $\mu_e = \mu_0 = 1.658$. Thus for the intermediate angles the effective value of μ_e lies between 1.486 and 1.658, so that when the angle of incidence at the end faces exceeds a certain limit, the canada balsam becomes optically less dense than calcite and E-ray is totally reflected by the balsam This limits the angle of incidence for the E-ray also to about 14°. Thus to avoid the transmission of O-ray and the total internal reflection of E-ray the angle between the extreme incident rays is limited to about 28°. Hence a *nicol prism cannot be used in highly convergent or divergent light.*

Nicol as an analyser. The function of the nicol as an analyser of plane-polarised light is easily understood. Here two nicols P and A are arranged co-axially and they constitute a polariscope ; the first prism is the *polariser* and the second the *analyser.* When the nicols are similarly orientated, that is, the principal section of the analyser *A* is parallel to the principal section of the polariser *P*, the extraordinary ray from the polariser is freely transmitted by the analyser. This position and the other position corresponding to the angle of 180° between the two principal sections when they are again parallel are referred to as "parallel nicols".

When the analyser *A* is turned through 90° from these positions, that is the principal sections of the two are mutually perpendicular, the nicols are said to be 'crossed' and no light is transmitted by the system. That this should be the case is obvious as the ray on emerging out of the polariser has vibrations in its principal section and, therefore, perpendicular to the principal section of the analyser. It will thus have no component in the principal section of the latter and will travel as an ordinary ray in it to be totally reflected at the balsam layer.

For intermediate positions the *E*-ray transmitted by the polariser on being incident at the analyser is broken up into two components one having vibrations parallel to the principal section of the analyser and the other perpendicular to it. It is the former component which passes through and the latter is ultimately reflected at the balsam layer. The intensity of the beam emerging from the polariscope obeys cosine square law, *viz.*, $I \propto \cos^2 \theta$, where θ is the angle of rotation from the position of maximum intensity.

Polarisation by selective absorption by crystals. The phenomenon of polarisation by *selective* absorption is shown by a number of substances, the most outstanding being tourmaline. It is a doubly refracting crystal which transmits partially only the *E*-rays whose vibrations are parallel to the optic axis of the crystal and absorbs completely the *O*-rays vibrating in a plane at right angles to the optic axis. Its absorbing power depends upon its thickness and a few mm thick crystal will easily absorb the *O*-rays. Such doubly refracting crystals are known as dichromic crystals and this property measured by the difference in the absorption of the O and E rays is called dichroism.

Polaroids

The phenomenon of selective absorption is made use of in the construction of polaroids. These are large-sized polarising films of any cross-section which are used to obtain plane polarised light easily. They can be made by depositing on thin sheets of nitrocellulose *ultra-microscopic polarising crystals* of an organic compound called herapathite (periodical of quinine sulphate), so called after the name of its discoverer Herapath. These tiny needle-like crystals are oriented by forcing them under pressure through a fine horizontal slit so that their optic axes are all parallel to each other and the film then forms one large single crystal which absorbs the O-rays and transmits the E-rays. Such polaroid films can serve as excellent substitute for nicol prism.

Uses of polaroids. These polaroids have several commercial possibilities besides their use in polarising instruments such as the study of the optical properties of metals, the analysts of crystals, the Control of light intensity, cutting off the glare produced by light reflected from polished metals. Another interesting application is the elimination of dazzle from head lamps and wind shields. The head-lights or wind shields are covered with polaroid discs so that the vibration plane of the transmitted light is at 45° to the horizontal in the anti-clockwise direction as seen by its own driver. This would enable him to see outside objects by the light from his lamps. An approaching car with head-lights similarly covered would give light polarised in a plane at right angles to the above so that the driver of the opposite car seeing it through a polaroid arranged with its vibration plane parallel to the vibration plane of its own head-lights would not receive the light and hence the dazzle would be eliminated.

Polarisation by scattering. We discussed the phenomenon of the scattering of light by small particles and showed that the blue colour of the sky is due to the shorter components of white light being more intensely scattered. Arago observed in 1811 that this scattered blue light is polarised in a plane containing the sun and the ray from the sky. If the solar light scattered along lines perpendicular to the path of the original beam is observed through a nicol prism, it is found that as the nicol is rotated about its axis, the intensity of light changed from a maximum to a minimum every 90°. This shows in a simple way that light is plane-polarised and this is what we should expect from the transverse nature of light waves. If the incident beam is horizontal, the vibration plane for light scattered in a perpendicular direction must be a vertical one, since the vibrations in both the incident and the scattered beams must be perpendicular to the direction of propagation of the waves. Hence, the vibrations in the beam scattered horizontally at right angles to the incident beam must be confined to a vertical plane. Now if the position of the nicol is noted when a maximum amount of light is transmitted, it is found that its principal plane is also vertical.

We arrive at the same conclusion by interposing the nicol in the path of the incident beam. If now we look in a direction at right angles to the beam with unaided eye we again observe variations in brightness as the nicol is rotated about its axis. Moreover, the principal plane of the nicol is vertical for maximum intensity and horizontal for minimum intensity, that is, there is almost complete polarisation in this direction. In all other directions the scattered light will be only partially polarised. We can easily explain these results if the vibration plane of the light transmitted by a nicol is a principal plane. Assuming the principal plane of the polarising nicol to be vertical the plane-polarised beam on being incident on the scattering particles causes vibrations in a vertical direction, that is, along lines at right angles to the line of sight. On the other hand if the principal plane is horizontal, vibrations are caused parallel to the line of sight and hence no disturbance in the line of sight. The reason, why the light scattered in a plane transverse to the direction of propagation has no component parallel to the direction of propagation of the incident light, is that the light being a transverse wave, can have no horizontal component.

Propagation of light in uniaxial crystals. We have seen that in double refraction, a single ray gives rise to two refracted rays, ordinary and extraordinary. It was Huygens who first of all examined the propagation of light in uni-axial crystals. He assumed that a wave which is propagated in a uniaxial crystal has two surfaces, *one within the other.* He found that while the wave-front of the ordinary ray which obeys the ordinary laws of refraction and travels with the same velocity in *all* directions in a crystal is a sphere and that of the extra-ordinary ray which does not obey the laws of refraction and travels with different velocities in different directions, is an ellipsoid. Since no double refraction occurs along the optic axis the surfaces of sphere and ellipsoid touch in two points, the lice joining which gives the direction of the optic axis of the crystal.

Negative and positive crystals. There are some crystals like calcite, in which the *E*-ray travels faster than the *O*-ray, the sphere is inside the ellipsoid. The diameter of the sphere is equal to the minor axis of the ellipsoid. Such crystals are known as negative crystals.

There are other crystals, like quartz, in which the *E*-ray travels slower than the *O*-ray, the sphere is outside the ellipsoid. The diameter of the sphere is equal to the major axis of the ellipsoid. They are called positive crystals.

Principal refractive indices. From the preceding sections we find that for any crystal the velocity of the ordinary ray is the same in all directions while that of the extra-ordinary ray is different in different directions but along the optic axis both have the same velocity. From this we conclude that in uniaxial crystals, there are two principal refractive indices, one corresponding to the velocity of the *O*-wave and the other to the velocity of the *E*-wave, when travelling normal to the optic axis. In all types of uniaxial crystals, the refractive index for the ordinary ray is defined as

$$\mu_0 = \frac{\textit{velocity of light in air}}{\textit{velocity of ordinary wave}}.$$

The principal index of extra-ordinary wave in negative uniaxial crystals like calcite is expressed as

$$\mu_0 = \frac{\textit{velocity of light in air}}{\textit{maximum normal velocity of E wave}}$$

While in positive uniaxial crystals like quartz the principal index for the extra-ordinary wave is expressed as

$$\mu_0 = \frac{\textit{velocity of light in air}}{\textit{minimum normal velocity of E wave}}$$

The maximum or minimum normal velocity of the E wave in either type is its velocity in a plane perpendicular to the optic axis.

Since the *E* wave touches the *O* wave along the optic axis both have the same velocity along the axis. Each pair of the values of μ_0 and μ_e, therefore, determines the ratio between the two axis, of the ellipsoidal extra-ordinary wave surfaces.

$$\frac{\mu_0}{\mu_e} = \frac{CD}{AB}$$

For *negative* crystals, V_e, the *maximum* normal velocity of the *E* wave is greater than V_0, the velocity of the ordinary wave and so $\mu_e < \mu_0$, and for *positive* crystals *V*., the *minimum* normal velocity of the *E* wave is greater than V_0, the velocity of the ordinary wave so that $\mu_e > \mu_0$ where μ_e and μ_0 are the refractive indices for the *O* and *E* waves respectively.

Huygens' Theory

Huygens extended his idea of ordinary refraction on the wave theory to the case of double refraction in uniaxial crystals. He assumed that each point of the crystal surface becomes the centre of two secondary wavelets which spread out into the medium (crystal). The spherical one corresponds to the ordinary wave while the ellipsoidal one to the extraordinary wave The minor or the major axis of the ellipsoid is parallel to the optic axis and the two waves have the same velocity along this direction. In any other direction *E* wave travels with different velocity and the difference is maximum for directions at right angles to the optic axis. Huygens' inferences have been confirmed by experiments. To illustrate this we shall consider below a few simple cases. It may be noted that in all the following cases *AB* represents the trace of an incident plane wave-front perpendicular to the plane of the paper. It is incident on the refracting face *XX'* of a doubly refracting negative crystal (*i.e.*, calcite). The face *XX'* is also perpendicular to the plane of the paper. The dotted line *AZ* represents the direction of the optic axis. The construction for the refracted wave-fronts is obtained by the application of the principles of wave theory of light discussed already, except that here, instead of one, there are two waves, one spherical and the other ellipsoidal.

Case 1. Optic axis lying in the plane of incidence and inclined to the refracting surface.

(i) *Oblique incidence.* *AB* is the incident plane wave-front which meets obliquely at the surface *XX'* of the crystal first at the point *A* and makes it the centre of ordinary and extra-ordinary wavelets which travel in the crystal with wave-fronts-as spherical and ellipsoidal respectively. During the time $t(=BC/V_a)$ in which the disturbance from *B* reaches *C*, V_a being the velocity of light in air, ordinary spherical wavelet has travelled a distance equal to

$$t \times V_0 = \frac{BC \times V_0}{V_a}$$

$$= \frac{BC}{\frac{V_a}{V_0}} = \frac{BC}{\mu_0}.$$

where V_0 is the velocity and μ_0 the refractive index for the ordinary ray. To find the position of the ordinary wave-front in the crystal draw a sphere with A as centre and BC/μ_0 as radius. From C draw a tangent plane CO touching the sphere at O. A plane passing through CO and perpendicular to the plane of the paper represents the position of the ordinary refracted wave-front. AO is the refracted ordinary ray which lies in the plane of incidence and obeys the laws of refraction. The value of the refractive index μ_0 is constant for all positions of the optic axis.

The extra-ordinary wavelet is an ellipsoid touching the ordinary wavelet at P, a point on AZ which represents the direction of the optic axis AP also represents the semi-minor axis of the ellipse-and is equal to BC/μ_0, since its refractive index along the optic axis has the maximum value and is equal to μ_0. At right angles to the optic axis the velocity of the E wave, V, is maximum and the value of its *principal refractive index is minimum.* Thus the semi-major axis of the ellipsoidal wavelet from A is given by

$$t \times V_e = \frac{BC}{V_a} \times V_e$$

$$= \frac{BC}{\frac{V_a}{V_e}} = \frac{BC}{\mu_e}$$

where V_e is the velocity and μ_e is the principal refractive index of the E-ray.

As in the case of the ordinary beam, the tangent plane CE

touching the ellipsoid at E represents the position of the extraordinary wave-front. It is a plane passing through C and E and perpendicular to the plane of the paper. AE is the direction of the extra-ordinary ray. Thus the O-ray and -E-ray travel along different directions with different velocities except along the optic axis when both the rays have the *same velocity* and *are unpolarised.* They emerge out without any phase difference, so to say, one is merged into the other. In this case the principal planes of O and E rays are coincident with the plane of incidence. In other directions both of them are plane polarised in mutually perpendicular planes. When they travel at right angles to the optic axis they follow the same path but possess different velocities, the difference being maximum there.

In a more general case when the optic axis is not in the plane of incidence, the point of contact of the tangent CE to the ellipsoidal surface will not lie in the plane and hence the extraordinary ray will be inclined to the plane of incidence.

(ii) *Normal incidence.* Draw tangent planes OO_1 and EE_1 respectively to the spherical and ellipsoidal surfaces originating from the centres A and B where the incident waves strike the crystal simultaneously. The planes through OO_1 and EE_1 perpendicular to the incident surface represent the ordinary and extraordinary wave fronts in the crystal.

Case-2. Optic axis lying in the plane of incidence and perpendicular to the refracting surface.

(i) *Oblique incidence.* In this case, the two wave-fronts, ordinary and extra-ordinary, touch each other at P. The line joining AP represents the direction of the optic axis in this case.

Proceeding in the same manner as in case 1 above the

planes passing through *CO* and *CE* and perpendicular to the plane of the paper respectively represent the wave-fronts of the *O* and *E* waves.

(ii) *Normal incidence.* *E* and *O* waves coincide at all instants and travel in the *same* direction with the *same* velocity along the optic axis. There is no double refraction in this case.

Case 3. Optic axis lying in the plane of incidence and parallel to the refracting face.

(i) *Oblique incidence.* In this case the optic axis lies along the line *XX′*. The two secondary surfaces, the sphere and the ellipsoid, touch on this line. The positions of the *O* and *E* wave-fronts, *CO* and *CE*, are drawn on exactly the same considerations as in case 1 above. The *O* and *E* refracted wave-fronts travel with different velocities along different directions.

(ii) *Normal incidence.* In this case OO_1 and EE_1 are the refracted wave-fronts at the same instant of time. They are parallel to each other and travel in the *same* direction (perpendicular to the refracting surface *XX′*) but with *different* velocities, thus producing a path difference between the *O* and *E* waves on emergence but there is no separation into two beams. This property is made use of in the construction of quarter- wave and half-wave plates.

Case 4. Optic axis perpendicular to the plane of incidence and parallel to the refracting surface.

(i) *Oblique incidence.* In this case sections of *O* and *E* wave-surfaces in the plane of incidence are both circles regardless of the inclination of the incident wave-front as the optic axis is perpendicular to this plane. The positions of *O* and *E* wave-fronts can be constructed in the same manner as in case 1 above. It is evident that

O and *E* waves travel with different velocities along different directions.

(ii) *Normal incidence.* In this case both the *O* and *E* waves travel in the *same* direction with *different* velocities but *no separation takes place.*

It may be noted that in this case extra-ordinary wave has the *same* velocity in *all* directions in the plane of incidence and so it obeys both the laws of refraction as faithfully as the ordinary wave does.

The construction of double refraction as described above has been experimentally verified on crystals cut differently with respect to the optic axis by Huygens, Mascart, Stokes and Glazebrook and has been found to agree with the theoretical results to within 1 30,000.

Determination of refractive indices. The principal may be determined experimentally indices for uniaxial crystals by using prisms cut in the manner.

In prism (*a*) the crystal is so cut that the optic axis is perpendicular to the base; and in the position of minimum deviation the *E*-ray is travelling almost perpendicular to the optic axis, principal index μ_s.

In prism (*b*) the optic axis is parallel to the refracting edge of the prism so that the cross-section of both the wave surfaces will be circles and Snell's law will be valid for all angles of incidence.

If either of these prisms is placed on the table of a spectrometer two images are observed and the angle of minimum deviation can be measured for each. The *O* and *E* indices are calculated in the usual way by the formula

$$\mu = \frac{\sin \dfrac{A+D}{2}}{\sin \dfrac{A}{2}},$$

where A is the angle of the prism and D is the angle of minimum deviation for either ray.

The principal indices for a number of doubly refracting substances are given in the table below:

Substance	μ_0	μ_e
Iceland spar (– ve)	1.6583	1.4864
Tourmaline (– ve)	1.640	1.620
Nitrate of soda (– ve)	1.585	1.337
Quartz (–ve)	1.544	1.553
Ice (+ve)	1.303	1.307

Doable image prisms. Nicol prism, described earlier, cannot be used in ultra-violet light because canada balsam is opaque to these shorter wavelengths. To overcome this difficulty other types of prisms, made of calcite or quartz and cemented together with glycerine or castor oil, have been designed. In this connection Rochon and Wollaston prisms need special mention. These optical devices, also known as double image prisms, are employed to obtain two images of au object near together. Mostly they are used in telescopes for finding the angular diameters of placets.

(1) *Rochon's prism.* It consists of two right-angled prisms *ABC* and *ADC* made of quartz with their retracting angles equal. They are cemented together to a rectangular block. The optic axis of the prism *ABC* is parallel to its base *BC* while that of the prism *ADC* is perpendicular to the plane of the paper. A ray incident normally on the face *AB* of the first prism passes through it without division being along the optic axis. On entering into the second prism *ADC* at *P* it is split up into *O* and *E* rays. *O*-ray is transmitted through it without any deviation while *E*-ray is refracted along

PE, that is, it is deviated towards the base of the prism *ADC*, since in quartz μ_0 is less than μ_s. The *O*-beam is achromatic and the *E*-beam is chromatic. The *E*-beam at sufficiently large distances from the prism can be easily got rid of the other. This prism is generally required in optical instruments where only one plane-polarised beam is needed. In this case light must enter from the left first, parallel to the optic axis of the first prism.

(2) *Wollaston's prism.* This is another double image prism known as Wollaston prism. It is similar in construction to the Rochon prism except that its first prism $A_1B_1C_1$ has its optic axis parallel to the face A_1B_1. The incident light in this case, enters normally to the surface A_1B_1 and travels at right angles to the optic axis. The two beams *O* and *E*, travel in the first prism in the *same direction* but with *different* speeds. When the rays strike the surface A_1C_1 of the second prism at P_1, they *change* roles, the ordinary ray is refracted as extra ordinary and the extraordinary ray as ordinary, since the principal-planes of the two prisms are at right angles. This deviates both the *O* and *E* rays so that there is a greater separation of the two chromatic beams. The refractive index increases for one ray and decreases by the same amount for the other ray so that the two rays are deflected in the opposite directions which show that they form two widely separated beams vibrating in perpendicular planes. This device is especially useful where the intensities in polarised light are involved.

10

Relative Optics

Einstein's Special Theory of Relativity

A major turning point in the Field theories proposed from time to time to interpret the physical nature of the universe was the formulation of the special theory of relativity in 1905 by Albert Einstein (1879-1955). It has emerged as a consequence of the failure of the attempts to discover the existence of the enigmatical luminiferous ether which was coined out of necessity to explain the propagation of light waves by Huygen and was imagined to fill all space and to be a perfectly elastic solid. Later in 1864 Maxwell gave his electromagnetic theory of radiation, which also postulated the existence of a medium endowed with the properties of permittivity and permeability and the same were considered as the attributes of Huyget's luminiferous ether. Assuming its existence therefore, it might be conjectured that the motion of the planets and stars would in some way be affected by such an all pervading medium through which they were moving, but astronomical observations, accumulated over centuries, indicated no such effect. Since all speculations about the existence of ether stem from its properties as a medium for the propagation of light waves, it was expected that a sensitive optical experiment to determine whether light waves could be influenced by the motion of the earth through the ether, might settle the issue. The object of such an experiment would be to discover the ether drift equivalent to the orbital velocity of the earth of 29.6 km/sec around the sun, assuming the ether to serve as a stationary frame of reference. It was to find the effects of this motion on the velocity of light that an experiment was designed by A. Michelson and H. Morley in 1887, which we shall now describe in detail.

The Michelson and Morley experiment. The object of the experiment was to test whether there was any relative motion between the earth and the luminiferous ether. The apparatus utilized was essentially a Michelson interferometer with its arm OM_2 parallel to the velocity of the earth around the sun. To avoid distortion of the instrument by strain the entire apparatus was mounted on a rigid platform floating on mercury in an iron trough and capable of slow rotation about an axis perpendicular to the plane of paper at a rate of ten complete rotations per hour. A beam of light originating at S is partly reflected at the back surface of A to M_1 and partly transmitted to M_2 ; M_1 and M_2 being optically equidistant from O, a point in the rear surface of A. The two beams are reflected from the front surfaces of M_1 and M_2 and meet at O, where they produce an interference pattern consisting of dark and bright band, visible to an observer at E. Let the length of each optical path be d.

Suppose the speed of light in a stationary ether is c and that of earth round the sun is v. The speed of light along OM_2 relative to the apparatus is $(c - v)$ and on return from M_2 to O its speed is $(c + v)$. The total time t, to go from O to M_2 and back to O is thus

$$t_1 = \frac{d}{c-v} + \frac{d}{c+d}$$

$$= 2d.\frac{c}{c^2 - v^2}$$

$$= \frac{2d}{c}\left[1 + \frac{v^2}{c^2} + \frac{v^4}{c^4} + \ldots\right]$$

$$= \frac{2d}{c} + \frac{2dv^2}{c^2} \text{ approx.} \qquad \ldots(i)$$

using binomial theorem and neglecting powers of v/c higher than two.

In going from O to M_1 and back to O, the actual path of the light beam through the ether is along $OM_1'O'$. This is due to the motion of the apparatus through the ether. The speed of light along this path is c so that its speed along the path OM_1 and M_1O is $(c^2 - v^2)^{1/2}$. The time taken by the light to go from O to M_1 and back to O is given by

$$t_2 = \frac{2d}{\left(c^2 - v^2\right)^{1/2}}$$

$$= \frac{2d}{c}\left[1 + \frac{v^2}{2c^2} + \ldots\right]$$

by binomial theorem

$$= \frac{2d}{c} + \frac{dv^2}{c^3} \text{ approx.} \qquad \ldots(ii)$$

Hence ΔT the difference in the times for the two paths is given by

$$\Delta T = t_1 - t_2 = \frac{2d}{c} + \frac{2dv^2}{c^3} - \frac{2d}{c} - \frac{dv^3}{c^3} = \frac{dv^3}{c^3}$$

There is therefore a path retardation of $\Delta T \times c = dv^2/c^2$ between the two paths.

After having noted the position of the interference fringes with OM_2 parallel to the motion of the earth the whole apparatus is rotated through 90°so that the path OM_1 becomes parallel to the direction of the earth's motion, *i.e., the two paths interchange.* Such a rotation makes the path OM_1 and back optically longer than OM_2 and back by dv^2/c^2 instead of being the shorter, thus introducing a total path retardation between the two interfering beams,

$$\delta = 2\Delta T \times c = \frac{2dv^2}{c^2}$$

Hence a displacement of interference fringes will occur. If f be the fraction of the fringe shift, then

$$f\lambda = \frac{2dv^2}{c^2}$$

or

$$f = \frac{2d}{\lambda}\left(\frac{v}{c}\right)^2 \qquad ...(iii)$$

Now from the known motion of the earth $\left(\frac{v}{c}\right)^2 = 10^{-8}$ approx.; d was taken equal to 11 m and $\lambda = 6 \times 10^{-7}$ m therefore we have $f = \frac{2 \times 11}{6 \times 10^{-7}} \times 10^{-8} = 0.37$ or about $\frac{1}{3}$ of a fringe width.

The actual displacement observed was certainly less than 20th part of this amount and probably less than the fortieth part. *Thus a diligent search failed to measure ether drift.* This means that the relative velocity between the earth and the ether is zero. The experiment has been repeated and other experiments have also been performed. The weight of evidence is in favour of the view that the existence of the ether cannot be proved by such experiments. Recently Dr. C.H. Townes of Columbia University (1958) used the *maser*, an instrument capable of measuring very short time intervals with very great accuracy to recheck the Michelson-Morley experiment. The results are in close agreement with that of Michelson and Morley, that is, the speed of light in empty space is constant and is not affected by the motion of the observer or the source through the ether. *We, therefore, conclude that ether does not exist.* This led Einstein to re-examine the fundamental laws of mechanics and he then proposed his theory of relativity which affords a satisfactory explanation of the experiment besides having the advantage of logical perfection and security of foundations, which are the essential attributes of a principal theory. Before we take up the postulates of the Einstein's Special Theory of Relativity and its consequences, let us briefly explain Newtonian relativity of inertial systems. (*An inertial system of reference in physics is one to which events are spatially referred and in which, when there are no forces, an object moves with constant rectilinear velocity*).

Newtonian Relativity

The student is familiar with Newton's three laws of motion. These laws hold good to a fair degree of approximation when referred to a system of axes relative to earth. Let us now investigate how these laws are modified in a frame of reference moving with a velocity v in the positive X-direction, say.

Let F $(OXYZ)$ be a fixed frame of reference and F' $(O'X'Y'Z')$ a frame moving with uniform velocity v in the positive X-direction so that O, O' coincide at time $t = 0$. Let each describe the rectangular coordinates of the same point P. According to an observer in frame F the coordinates of P at time t are x, y, z, and according to that in frame F' they are y', y', z'.

The necessary equations of transformation for the frame F, in motion relative to the fixed frame F are

$$x' = x - vt;$$

$$y' = y;\ z' = z \quad \text{... (1)}$$

Differentiating these equations with respect to time, we get

$$\dot{x}' = \dot{x} - v\ ;\ \dot{y}' = \dot{y}'\ ;\ \dot{z}' = \dot{z} \quad \text{...(2)}$$

This is described as the *classical transformation for non-accelerated systems.* When critically examined it implies that space and time are absolute entities independent of each other and independent of the frame.

These equations refer to the components of the velocity of the particle as determined by observers in the two coordinate frames.

The corresponding equations of acceleration in the two frames are obtained by a second differentiation and are

$$\ddot{x}' = \ddot{x} ; \ddot{y}' ; \ddot{z}' = \ddot{z} \qquad \dots (3)$$

It follows that the acceleration of the particle is the same in the two frames and hence Newton's second law of motion $F=ma$, is equally valid when transformed to a set of axes moving with uniform velocity with respect to the original one or in other words, the Newton's laws of motion are *invariant, i.e. unaltered in form by a change of axes,* under the so-called Galilean-Newtonian type of transformation.

This obviously leads to the conclusion that an observer within an inertial moving system cannot detect his motion by any dynamical experiments performed wholly within that moving system. This result is called Newtonian or classical relativity and classical dynamics confirms it. Further, it may be noted that 'classical dynamics' search [Michelson-Morley experiment] for a fixed frame of reference in stationary ether has failed. The wave theory of light and Maxwell's electrodynamics sought luminiferous ether as a unique reference frame but all attempts to establish that frame were fruitless. Both experiment and theory show that if electromagnetic measurements stated with respect to fixed axes are transformed to *axes moving with velocity v in the X-direction,* the form of transformation rule changes. In order to maintain *a principle of relativity* it is necessary to be able to transform from axes at rest to axes in motion *without changing the forms of the equations of motion.* That is, Maxwell's equations of electromagnetic field should remain *invariant.* But this is not so. All these failures demanded a new orientation of the relativity which would include dynamical as well as electromagnetic phenomena.

Fundamental postulates of Einstein's special theory of relativity. In attempting to explain the results of Michelson-Morley experiment and other similar experiments to detect motion through the ether, Einstein in 1905 arrived at a new idea of assuming an ether filling empty space as a *meaningless concept* and that only motion *relative* to material bodies has physical significance. Accordingly with a view to harmonise this assumption with the known laws of optics and electromagnetism, he based his new theory which is known as special or restricted theory of relativity on two fundamental postulates :

(1) The *first postulate* is based on classical theory and States that 'the fundamental physical laws should have the same mathematical form in all inertial systems in uniform translatory motion with respect to one another.' This is another way of stating the requirement that the equations of a physical theory be *invariant* with respect to coordinate systems moving with different velocities. This is the *principle of equivalence.*

(2) The *second postulate* is based on the results of Michelson-Morley experiment and other optical and electrodynamic measurements. It may be stated as follows :

The speed of light in empty space is constant and is independent of the source and the observer. No signal or energy can be transmitted with a speed greater than the speed of light.

This means that a light wave leaving a star will have a

constant velocity of 3×10^5 km per second relative to an observer regardless of the fact whether he and the star are approaching or receding from each other with any velocity, whatsoever.

The speed of light is defined as the distance traversed by the beam of light divided by the time required to traverse this distance as measured by any observer in a given reference frame.

We give below Lorentz-Einstein transformation based on the above postulates. It would enable us to interpret the dynamical and electromagnetic phenomena and to derive the consequences of the special theory of relativity.

Lorentz-Einstein Transformation

If the postulates of Einstein's relativity are accepted we shall have to formulate equations of transformation expressing the relation between measurements made by observers in different frames of reference so as to enable us to transform the coordinates of one frame into those of other. Let us consider again a reference frame F' ($O'X'Y'Z'$) moving relative to frame $F(OXYZ)$ with uniform velocity v in the positive X-direction.

Suppose an event observed in the F frame occurs at coordinates x, y, z at time t, the same event observed in the F' frame has coordinates x', y', z' and will be observed at a time t', the two instants t and t' must not necessarily be the same even though the clocks used by the two observers stationed in them are identical in all respects and have been properly synchronised at the instant they passed each other at the origin of coordinates. Thus to characterise an event we require in addition to the three space coordinates x, y, z, the time t as the *fourth* coordinate. This mean*! that space and time are not separate entities but are inextricably bound to each other forming a four dimensional space-time continuum so that one without the other becomes meaningless.

To derive the equations of transformation of space and time variables appropriate to the special theory of relativity we choose a physical event, say, the emission of a pulse of light from a source and consider the description of this event by the two observers subject to two important limitations : *(i) The measuring instruments in the two frames cannot communicate directly except when they coincide in position.* (*ii*) *The only means of communication between the two frames is a pulse of light.* Suppose the source of light is at O, the origin of coordinates of frame F and suppose that O' the origin of coordinates of frame F' coincides with O at the instant the pulse of light is emitted and let $t = t' = 0$ at this instant. According to second postulate of relativity the speed of light is constant and independent of the motion of the observer, hence each one will note a spherical wave spreading out from his origin of coordinates. If c is the speed of light, the position of the spherical wave will be given by

$$x^2 + y^2 + z^2 = c^2t^2 \quad ...(4)$$

in the F frame,

and by $x'^2 + y'^2 + z'^2 = c^2t'^2$ ...(5)

in the F' frame.

$$\text{Thus,} \quad \frac{x^2 + y^2 + z^2}{t^2} = \frac{x'^3 + y'^2 + z'^2}{t'^2} = c^2 \quad ..(6)$$

(constant) for either frame.

To find the relation between x and x', t and t' we use the Galilean transformation as a guide and assume that the equations will be *linear* and of the forms,

$$\left.\begin{aligned} x' &= K(x - vt) && ...(a) \\ y' &= y && ...(b) \\ z' &= z && ...(c) \\ t' &= At + Bx && ...(d) \end{aligned}\right\} \quad ...(7)$$

where the constants, K, A and B are to be determined.

Substituting set of equations (7) in equation (5) we have

$K^2 (x - vt)^2 + y^2 + z^2 = c^2 (At + Bx)^2$

Expanding the above equation and collecting terms gives

$(K^2 - B^2c^2)x^2 + y^2 + z^2 - (A^3C^2 - K^2v^2)t^2 - 2(ABc^2 + K^2v)\, xt = 0$

Now $x^2 + y^2 + z^2 - c^2 t^2 = 0$ by equation (4)

Comparing coefficients of like terms in them, we get

$$K^2 - B^2C^2 = 1 \qquad ...(i)$$

$$A^2C^2 - K^2v^2 = c^2 \qquad ...(ii)$$

$$ABC^2 - K^2v = 0 \qquad ...(iii)$$

The solution of these equations yields

$$K = A = \left(a - \frac{v^2}{c^2}\right)^{-\frac{1}{2}}$$

and

$$B = \frac{Kv}{v^2},$$

so that the equations of transformation become

$$\left.\begin{aligned} x' &= K(x - vt) && ...(a) \\ y' &= y && ...(b) \\ z' &= z && ...(c) \\ t' &= K\left(t - \frac{vx}{c^2}\right) && ...(d) \\ \text{with} \quad K &= \left(1 - \frac{v^2}{c^2}\right)^{-\frac{1}{2}} && ...(e) \end{aligned}\right\} \quad ... (8)$$

These equations are known as Lorentz-Einstein equations of transformation of space and time coordinates and enable us to transform the coordinates of the moving frame $(x'_t\, y', z', t')$ into those of the stationary frame (x, y, z, t).

The transformation which is inverse of (8) can be obtained by replacing v by $-v$.

$$\left.\begin{aligned} x' &= K(x + vt) && \quad ...(a) \\ y' &= y && \quad ...(b) \\ z' &= z && \quad ...(c) \\ t' &= K\left(t' + \frac{vx'}{c^2}\right) && \quad ...(d) \end{aligned}\right\} \quad ...(9)$$

This follows immediately from the fact that the frame F is moving with a velocity $-v$ relative to frame F'. The value of K however remains unchanged. It may be noted that in both these sets of equations the lengths perpendicular to the direction of motion are not affected. If the motion of the moving frame were not along the x-axis exclusively, then y', z', coordinates would also have transformation equations similar to those for x'.

Lorentz found that the results for a system moving with velocity v in the X-direction are simplified considerably when they were expressed as a function of $\left(t - \frac{vx}{c^2}\right)$ as derived above instead of as a function of t. Hence $\left(t - \frac{vx}{c^2}\right)$ may be regarded as a sort of local time for the plane specified by X. According to the *principle of relatively* then the earth's motion has no influence, whatsoever, upon optical phenomena, because each observer carries with him his own system of space and time and the units which he employs to measure space and time adjust automatically as to eliminate the influence of motion.

Maxwell's equations of electromagnetic field are invariant with respect to this transformation. They have the same form as for a system at rest, that it, if equations are expressed in terms of x, y, z, t and x', y', z', t' are substituted by means of the formulae above, the form of the equations is unaltered, since Maxwell's equations sum up all optical and electrical phenomena. All physical phenomena are supposed to be essentially electrical and they are hence unaffected by motion.

The distinctive feature of the new hypothesis is that it is quantitative and Einstein showed that it is necessary. There are certain phenomena, which cannot be explained on the assumption that space and time are absolute and hence that assumption must go. Relativity consequently is now accepted as a faith. It is inadvisable to devote attention to its paradoxical aspects.

Some consequences of the special theory. Let us now see how the fundamental concepts of length, time and mass are modified as a consequence of Lorentz-Einstein equations of transformation.

(i) *Relativity of length (Contraction of space).* From Lorentz-Einstein equations of transformation it can be shown that length or space is not absolute but *relative*. Suppose a rigid rod of length L_2 is lying in the frame F' and the axis x' is taken parallel to the length of the rod. Then in that frame the length of the rod can be expressed as

$$L_0 = x_2' - x_1', \qquad \text{...(10)}$$

where x_1' and x_2' are the coordinates of the two ends of the rod.

The length of the rod in the moving frame F' as measured by an observer in the fixed frame F would be the distance between the coordinates x_1 and x_2 of the ends of the rod at a given time t, so that

$$L = x_2 - x_1 \qquad \text{...(11)}$$

Using equation 8(a) we find that

$$\begin{aligned} L_0 &= x_2' - x_1' \\ &= K(x_2 - vt) - K(x_1 - vt) \\ &= K(x_2 - x_1) = KL \end{aligned}$$

from which

$$L = \frac{L_o}{K} = L_0 \left(1 - \frac{v^2}{c^2}\right)^{\frac{1}{2}} \qquad ...(12)$$

Thus the measured value of the length of the rod will be less when it is moving parallel to its length than when it is at rest with respect to an observer, the ratio of shortening being $\sqrt{1 - v^2 / c^2}$. If v has the value c, *i.e.*, if the matter moves with the velocity of lights *its length becomes zero.* This is clearly not possible, hence a particle can approach but never attains the velocity of light. The conctraction of an equivalent magnitude when physical objects move through the ether was originally postulated by Lorentz-Fitzgerald contraction hypothesis based upon electromagnetic conceptions which were considered justified because of the electrical theory of matter. It has recently (1959) been shown by Terrell that there is no substance in the contraction hypothesis. *It is only an apparent contraction and is inherent in the process of measurement thus accounting for the negative result of Michelson-Morley experiment.*

The apparent distortion of a moving object when seen by a stationary observer is due to the fact that the light forming the image is emitted at different times and from different positions of the object in space. This is just sufficient to cancel the Lorentz contraction. It may further be noted that in the derivation of equation (12) it is assumed that the light from the two ends of the rod is emitted and recorded simultaneously by the measuring instruments.

(ii) *Relativity of time (Dilation of time).* Let us now determine the relative values of time intervals as measured by the observers in two frames of reference in uniform relative motion. The equations 8(*d*) and 9(*d*),

$$t' = K\left(t - \frac{vx}{c^2}\right)$$

and $$t = K\left(t' + \frac{vx'}{c^2}\right)$$ respectively

clearly show that the measurement of time and time intervals by two different observers in uniform relative motion with respect to one another will depend upon this motion even though the same or identical clocks are used in their measurements.

Clocks moving with respect to an observer appear to go slow than they do when at rest with respect to him. If an observer in the frame F observes by his clock time interval Δt some event requires in the frame F', then his time interval will be longer than $\Delta t'$ determined by a clock in the moving frame F'. This effect is called *time dilation*.

To understand how time dilation comes about let us consider a clock at the point x' in the moving frame F'. When an observer in frame F' finds that time is t'_1, one in frame F will find it t_1, whence from equation 9 (*d*)

$$t_1 = K\left(t_1' + \frac{vx'}{c^2}\right)$$

After a time interval $\Delta t'$ (to him), the observer in the moving frame finds that the time is now t_2' according to his clock such that

$$\Delta t' = t_2' - t_1' \qquad ...(13)$$

The observer in F, however, measures the end of the same time interval to be

$$t_2 = K\left(t_2' + \frac{vx'}{c^2}\right)$$

So to him the duration of time interval Δt is

$$\begin{aligned} \Delta t &= t_2 - t_1 \\ &= K\,(t_2' - t_1')' \\ &= K\,\Delta t' \end{aligned}$$

to that $\Delta t > \Delta t'$. ...(14)

since $K = \left(1 - \dfrac{v^2}{c^2}\right)^{-1/2}$

A stationary clock thus measures a longer time interval between events occurring in a moving frame of reference than does a clock in the moving frame.

In other words, a clock moving with velocity v relative to an observer in frame F will lag behind one at rest in F' in the ratio $\sqrt{1 - v^2/c^2} : 1$. It always appears to go slow to an observer in F than it does to an observer at rest in its own frame *viz.*, the time processes are slowed down resulting in the dilation of time. *This opinion is reciprocal* for a clock at the origin of the frame F' where it obeys the relationship $\Delta t' = K\Delta t$ leading to a similar conclusion. This may be difficult to understand in an intuitive way ; it may take some time before you grasp the exact implication of this concept. The root of apparent paradox is the *invariance* of c. We see from the equations that if the moving frame is proceeding with the velocity of light that is $v = c$, then any time interval $\Delta t'$ occurring in it will be measured as Δt = infinity in the stationary frame-that is, the moving clock will appear to have stopped.

The effect must hold for any type of clock. In particular if T be the decay half-life of mesons or radioactive matter as measured in the frame F in which the particles are at rest, then

$T' = T/\left(1 - \dfrac{v^2}{c^2}\right)^{\frac{1}{2}}$ is the decay half-life observed in a frame F'

in which the particles are moving with velocity v. This predicted time has been confirmed by the experiments on the measurement of life-time of π mesons carried out by Durbin, Loar and Havens (1952). The π^+ mesons in a frame in which they are at rest have a mean life before decaying of about

2.5×10^{-2} sec. If the relativistic time dilation effect did not exist then assuming their velocity nearly equal to *c*, they would traverse a mean distance equal to $2.5 \times 10^{-8} \times 3 \times 10^{8}$ m/ sec = 75 m, before decaying. Actually they have been found to travel much farther than this because of time dilation, traversing sometimes a distance about 100 times the distance they would traverse before decay without the time dilation effect. The experiments on high energy particles generated in nuclear disintegration processes provide the physicist with tests for special relativity everyday.

We shall illustrate the time dilation of equation (14) by an example pertaining to the decay of μ mesons*.

Example. *The mean life of a muon at rest is about 2×10^{-6} s. If muons in cosmic rays have an average speed of 0.998 c, find the mean life of these muons and approximate distance they travel before decaying.*

Let y_0 be the altitude from ground at which a μ meson is produced. Its mean life $\Delta t'$ which is 2×10^{-6} seconds in its own *reference frame* will be extended in *our reference frame* on account of relative motion, to the value

$$\Delta t = K\Delta t'$$

$$= \frac{\Delta t'}{\sqrt{1 - v^2/c^2}}$$

$$= \frac{2 \times 10^{-5}\ \text{s}}{\sqrt{1 - (0.998\, c/c)^2}}$$

$$= \frac{2 \times 10^{-5}}{0.063}\ \text{s}$$

$$= \mathbf{32 \times 10^{-6}\ s}$$

almost 16 times greater than when it is at rest with respect to us. In 32×10^{-6} second a meson whose average speed is 0.998 c can travel a distance

$$y_0 = vt$$
$$= 0.998 \times 3 \times 10^8 \text{ ms}^{-1} \times 32 \times 10^{-6} \text{ s}$$
$$= 9500 \text{ m}$$
$$= 9.5 \text{ km.}$$

This explains how, despite their brief life spans, it is possible for the mesons to reach the ground from the high altitudes at which they are actually produced.

On the other hand, in classical physics time flows on *uniformly* for all observers on the same frame or on different frames relatively in motion without regard to anything external. Thus there was supposed to be a single time scale valid everywhere, whereas in relativity each observer has its own local time, which may be quite different from the times of other observers, *i.e.*, $t \neq t'$. The Newtonian conception of *'absolute'* time fails and the notion of *'simultaneity of events'* in respect of two frames relatively in uniform motion becomes untenable.

(iii) *Relativity of simultaneity*. From equations (8) and (9) we see that two events which are simultaneous in the frame F but occur at different places are not simultaneous in the frame F'.

For if $t_1 = t_2$

$$t_1' = K\left(t_1 - \frac{v}{c_2}x_1\right)$$

and
$$t_2' = K\left(t_1 - \frac{v}{c_2}x_2\right) \qquad ...(15)$$

where
$$K = \left(1 - \frac{v^2}{c^2}\right)^{-1/2}$$

Thus t_1' and t_2' are not equal if $x_1 \neq x_2$.

We shall Illustrate the relativity of simultaneity by an idealized.

Suppose a fast moving train 21.6×10^5 km long is travelling in a straight line with a uniform speed of 2.4×10^5 km/sec past a stationary observer *O* beside the railway track.

The train has two doors *A* and *B* one at either end which open on the arrival of a light flash. It is so arranged that they are visible to an observer *P* seated at the mid-point of the train and also to the stationary observer *O, P* sends a light flash just when he is opposite *O* and sees that both doors *A* and *B* open simultaneously

$$\frac{10.8 \times 10^5 \text{ km}}{3 \times 19^5 \text{ km / sec}} = 3.6 \text{ sec}$$

after the flash was sent (3×10^6 km per sec. is the speed of light). This is in accord with Michelson's celebrated experiment regarding the constancy of the speed of light.

The stationary observer *O* however, sees that the rear door B opens,

$$\frac{10.8 \times 10^5 \; km}{(3 \times 2.4) \times 10^5 km/sec} = 2 \text{ secs}$$

after the flash was sent, since the rear of the train moves at 2.4×10^5 km per second to meet the flash travelling at 3×10^6 km/ sec. But the front door *A* is moving forwards at 2.4×10^5 km/sec and the light flash travelling at 3×10^5 km/sec has to catch it up. It will, therefore, take

$$\frac{10.8 \times 10^5 \text{ km}}{(3 - 2.4) \times 10^5 \text{km / sec}} = 18 \text{ secs}$$

for the front door *A* to open, that is, for the stationary observer *O*, the front door *A* will open 16 sec later than the other door *B* opens.

Thus the two identical events occur simultaneously for *P* but at an interval of 16 secs for *O*. The relativity of simultaneity thus abolishes the idea of "absulate time", *i.e.*, time lapses at the same rate everywhere in the universe. In other words, the moment of occurrence of an event anywhere in the universe was the same everywhere else, however, the relativity of simultaneity has established that there is no such thing as same time for events as observed by observers in different frames of reference.

Note. In view of contraction of the speeding train the actual interval between the opening of the rear and front doors according to observer O will be 9.6 seconds and not 16 seconds, calculate how ?

(iv) *Relativity of velocity.* The Lorentz transformation leads to important formulas for the transformation of velocities from one frame to another. Suppose a particle is moving with a velocity w in the frame F' parallel to the *positive* direction of X -axis in the $X'Y'$ plane. According to classical relativity the velocity u_m of this particle relative to the stationary frame F is given by the well known equation

$$u_x = v + w, \qquad \text{...(16)}$$

where v is the uniform velocity of F' relative to F. We will now see that this relation does not hold when v is comparable to c.

The position of the particle at any time t' in the frame F' is given by the usual equation

$$x' = wt' \qquad \text{...(17)}$$

assuming that the particle starts at the Y'-axis at time $t' = 0$. Now the Lorentz-Einstein equations of transformation give

$$x' = K(x - vt) \qquad \textit{cf. } \text{equation } 8(a)$$

$$t' = K\left(t - \frac{vx}{c^2}\right) \qquad \textit{cf. } \text{equation } 8(d)$$

Substituting the values of x' and t' in equation (17), we have

$$K(x - vt) = wK\left(t - \frac{vx}{c^2}\right)$$

from which

$$x = vt + wt - \frac{wvx}{c^2}$$

or $$x + \frac{wvx}{c^2} = (v + w)t$$

or $$x\left[1 + \frac{wv}{c^2}\right] = (v + w)t$$

so that $$x = \frac{v + w}{\left(1 + \frac{wv}{c^2}\right)} t$$

therefore $$u_x = \frac{dx}{dt}$$

$$= \frac{v + w}{\left(1 + \frac{wv}{c^2}\right)} \qquad \text{...[18(a)]}$$

If w is toward – ve X-direction

$$u_x = \frac{v - w}{\left(1 - \frac{wv}{c^2}\right)} \qquad \text{...[18(b)]}$$

Also $$u_y = 0$$

and $$u_z = 0$$

Equations 18 (*a*) and 18 (*b*) give the relativistic rule for transforming parallel velocities.

Or more generally,

If w be the velocity in any random direction in $X'Y'Z'$ frame, then the relativistic velocities in both Y' and Z' directions will also be affected and we can deduce that

$$u_y = \frac{y}{t} = \frac{y'}{K\left(t' + \frac{vx'}{c^2}\right)} \quad \textit{cf.} \text{ equations [9(}b\text{) and (}d\text{)]}$$

$$= \frac{y'/t'}{K\left(1 + \frac{v}{c^2} \cdot \frac{x'}{t'}\right)} = \frac{w_y}{K\left(1 + \frac{vwx_\infty}{c^2}\right)}$$

$$= \frac{w_y\left(1 - \frac{y^2}{c^2}\right)^{\frac{1}{2}}}{1 + \left(\frac{vw_\infty}{c^2}\right)} \qquad \text{...[18 (}c\text{)]}$$

Similarly,

$$u_z = \frac{w_z\left(1 - \frac{v^2}{c^2}\right)^{\frac{1}{2}}}{\left(1 + \frac{vw_z}{c^2}\right)} \qquad \text{...[18(}d\text{)}$$

and from equation [18(*a*)] $u_x = \dfrac{v + w_x}{\left(1 + \frac{vw_x}{c^2}\right)}$, where $w_{x'}$, $w_{y'}$, $w_{z'}$ are respectively components in X', Y' and Z' directions in the frame F'.

Also note that equations 18 (*a*) and 18 (*b*) reduce to the Newtonian form $u_{x=} v \pm w$, when v and w are small as compared with c in which case wv/c^2 is negligible in comparison with unity.

In the special case in which $w = \pm c$, we have

$$u_x = \frac{v \pm c}{\left(1 \pm \frac{vc}{c^2}\right)}$$

so that $u_x = \pm c.$ ...(19)

That is, *the observer in F also measures the velocity of the particle as the velocity of light.*

This result should not surprise us as it confirms the second postulate of Einstein's relativity that the velocity of light is constant and is not affected by the motion of the source and the observer.

It is clear that *maximum velocity ever attainable is that of light,* for the addition of the velocity v to a particle with the velocity of light c results in a final velocity c. This limit to the velocity attainable in nature is of fundamental importance.

An important problem. Consider a particle moving with constant velocity along x-axis in frame F. Let us find its velocity from the point of view of two observers in the frames F and F' assuming that F is stationary and F' is moving relatively to it with a velocity v along the x-axis. According to observer in F' the particle moves from the point x_1' at time t_1' to the point x_2' at time t_2'. He then ascribes to it a velocity

$$v_1' = (x_2' - x_1')/(t_2' - t_1')$$

In a similar manner the velocity of the same particle is estimated by the observer in F to be

$$v_1' = (x_2 - x_1)/(t_2 - t_1)$$

Substituting from the transformation equation (9), for x_1, x_2, t_1, t_2 in the latter expression (remembering that v is the *relative* velocity of the two observers) leads to

$$v_1 = \frac{K\left(x_2' + vt_2'\right) - K\left(x_1' + vt_1'\right)}{K\left(t_2' + vx_2'/c^2\right) - K\left(t_1' + vx_1'/c^2\right)}$$

Dividing throughout by K and then rearranging gives

$$v_1 = \frac{\left(x_2' - x_1'\right) + \left(t_2' - t_1'\right)v}{\left(t_2' - t_1'\right) + v\left(x_2' - x_1'\right)/c^2}$$

Dividing top and bottom by $(t_2' - t_1')$ yields

$$v_1 = \frac{\left[(x_2' - x_1')/(t_2' - t_1')\right] + v}{1 + \left\{v/c^2\right\}\left\{(x_2' - x_1')/(t_2' - t_1')\right\}}$$

or $$v_1 = \frac{v + v_1'}{1 + vv_1'/c^2} \qquad ...(20)$$

This is the *law of addition of velocities,* and may be compared with the classical or Newtonian law, which states simply that if reference frames have a relative velocity v and if particle velocity estimated by F' is v_1' then that observed by F is just $v_1 = (v + v_1')$ since the denominator of the relativistic expression above is more than one the velocity measure in the stationary frame is less than $(v + v_1')$. So as observed from rest, moving objects in a moving frame are relatively slowed.

If the observer in the frame F' were to observe the phenomenon in the frame F and wishes to express v' in terms of *his* position and time system, he would have to use the transformation equation (8). If the substitutions are made, the result is, as the reader may verify,

$$v_1' = (v - v) / (1 - vv_1/c^2). \qquad ...(21)$$

Example. *Suppose two electrons in a stationary frame F move off with equal speeds of magnitude 0.9 c, one towards –X and other towards +X. What is the speed of one electron relative to the other?*

According to Galilean transformation their speed relative to each other measured in frame F is $0.9c - (-0.9c) = 1.8c$. This exceeds c.

But if we take $w = -0.9c$ so that frame F' keeps up with the electron going toward $-X$, then by equation 18 (*b*), the velocity of second electron relative to first, measured now in F', is

$$\frac{\left[0.9c - (-0.9c)\right]}{\left(1 + 0.9^2\, c^2/c^2\right)} = \frac{1.8}{1.81}\, c = \mathbf{0.9994\ c}$$

which is a little less than c.

For *alternative* solution of this problem see example 25-8.

Example. *The components of the velocity of a particle in a frame F' are $x_a = x_y = 0.6c$. This is observed from a reference frame F which moves along the X-axis with speed – 0.8c. Derive the velocity components of the particle measured from frame F.*

Using equation 18 (*b*) we have

$$u_x = \frac{0.6c-(-0.8c)}{1+\frac{0.6\times 0.8c^2}{c^2}} = \frac{1.4}{1.48}\ \text{c} = \mathbf{0.94\ c}$$

$$u_y = \frac{w_y\left(1-\frac{v^2}{c^2}\right)^{1/2}}{1+\frac{vw_x}{c^2}} = \frac{0{,}6c\,(1-0.64)^{1/2}}{1+\frac{0.6\times 0.8\,c^2}{c^2}}$$

$$= \frac{0.6\times 0.6}{1.48}\ c = \mathbf{0.24\ c}$$

and $\qquad u_z = 0.$

Note that although the relative motion of the two frames is parallel to X-axis, the velocity along Y direction is also affected. The slope of the path in AT plane is 1.0 in frame F' but becomes 0.24/94=0.25 in frame F.

Example. *A velocity vector has components $u_x = dx/dt$, $u_y = dy / dt$ and $u_z = dz/dt$ in the S-system, and $u_x' = d_x' = d_x' /dt$, $u_y = d_y' / dt'$ and $u_z = dz'/dt'$* in the moving S′-system. Show that

$$u'_x = (u_x - v) \Big/ \left(1-\frac{vu_x}{c^2}\right)$$

$$u_y' = u_y\left(1-\frac{v^2}{c^2}\right)^{1/2} \Big/ \left(1-\frac{vu_x}{c^2}\right)$$

and $$u_z' = u_z\left(1-\frac{v^2}{c^2}\right)^{1/2} \Big/ \left(1-\frac{vu_x}{c^2}\right)$$

Find the equations for the inverse transformation.

Putting equation (8) in the differential form

$$dx' = K\,(dx - vdt)\ ;\ dy; = dy\ ;\ dz' = dz$$

and $$dt' = K\,(dt - vdx/c^2)$$

From these expressions by substitution, we have

$$u_x' = \frac{dx'}{dt'} = \frac{K(dx - vdt)}{K\left(dt - \frac{vdx}{c^2}\right)} - \frac{u_x - v}{1 - \frac{vu_x}{c^2}}$$

$$u_y' = \frac{dy'}{dt'} = \frac{dy}{K\left(d' - \frac{vdx}{c^2}\right)} = \frac{dy/dt}{K\left(1 - \frac{vdx}{c^2\,dt}\right)}$$

$$= \frac{dy}{K\left(1 - \frac{vux}{c^2}\right)} = \frac{u_y\left(1 - v^2/c^2\right)^{1/2}}{1 - \frac{vu_x}{c^2}}$$

Similarly,

$$u_y' = \frac{dz'}{dt'} = \frac{u_z\left(1 - v^2/c^2\right)^{1/2}}{1 - \frac{vu_x}{c^2}}$$

The inverse equations can be obtained by using equation (9) in differential form.

(v) ***Variation of mass with velocity.*** Another important consequence of the special theory of relativity is that the masses of a particle varies with velocity and is given by

$$m = Km_0 = \frac{m_0}{\left(1 - \frac{v^2}{c^2}\right)^{1/2}} \qquad ...(22)$$

in which m_0 is the mass of the particle when at rest with respect to an observer and m is the mass when it is moving with velocity v relative to it. This is in contradiction to the Newtonian theory of invariability of mass, *i.e.*, the mass of a body does not depend upon its velocity. Since the factor $\sqrt{1 - v^2 / c^2}$ in equation (22) is always less than unity, we find that mass increases with velocity. The increase in mass is about $\frac{1}{2}$ % upto a velocity of l/10th that of light but rises much more rapidly with greater velocities being infinite when the velocity of light is reached. This relation has been experimentally confirmed by measurement of the masses of high speed electrons and β rays emitted by some radioactive substances by Bucherer and several other experimenters. The change of mass with velocity is very important in spectroscopic theory, for in the case of elliptic orbits the velocities of the electrons are so great when close to the nucleus that the mass change with velocity affects the energies of orbits considerably.

This fundamental result can be derived theoretically in several different ways, one of them being that in which we assume the principle of 'conservation of momentum' which is valid in all inertial frames of reference. But here we shall content ourselves by giving the expression only.

Relativistically therefore, we shall define the momentum p of a particle as $p = mv$, when m is the relativistic mass. Accordingly

$$P = \frac{m_0 v}{\sqrt{1 - v^2 / c^2}} \qquad \text{...(23)}$$

since $$m = \frac{m_0}{\sqrt{1 - v^2 / c^2}}$$

Table given below may be of interest to the student as it gives the value that the mass of an electron attains at different energies and velocities. The first and third columns, of course, would be the same for bodies of any *rest mass*.

Table showing relativistic mass increase of the electron

Velocity percent C	*Energy × 10^{-4} ev :*	*Mass*
0	0	m_0
86	0.5	$2m_0$
94	1.0	$3m_0$
96	1.5	$4m_0$
98	2.0	$5m_0$
99.5	4.5	$10m_0$
99.8	49.5	$100m_0$
100	∞	∞

The relationship between mass and energy of a relativistic particle. A particle of rest mass m_0 moving with a velocity v comparable with the velocity of light c is termed a *'relativistic particle.* In determining the kinetic energy of such a particle, we must take into account the mass variation with increase of velocity. This leads us to a modification of our ideas about energy. Suppose a force F acts parallel to the displacement ds of the particle, then the work done on it or the increase in kinetic energy is given by

$$dE_x = Fds$$

In classical physics force is defined as the time rate of change of momentum, so that

$$F = \frac{d}{dt}(mv) \qquad ...(24)$$

Therefore the change in kinetic energy,

$$dF_x = \left[\frac{d}{dt}(mv)\right] ds$$

$$= \frac{ds}{dt} \cdot d(mv)$$

But since $\frac{ds}{dt} = v$, we can write this equation as

$$dE_k = vd\,(mv)$$

$$= v^2dm + mvdv,$$

since mass is also variable.

This expression can be simplified by using the relativity mass transformation equation (22) according to which

$$m = m_0 \Big/ \left(1 - \frac{v^2}{c^2}\right)^{1/2}.$$

which after squaring and rearranging becomes

$$m^2c^2 = m^2v^2 + m_0^2\,c^2.$$

Differentiating this and noting that m_0 and c are constants we obtain

$$2mc^2dm = 3mv^2dm + 2m^2vdv$$

$$c^2dm = v^2dm + mvdv.$$

The right-hand side becomes identical with our expression for dE_k given above, so we have finally

$$dE_k = c^2dm. \qquad ...(25)$$

Thus a change in kinetic energy is *directly* proportional to the change in mass, the proportionality factor being the square of the velocity of light.

This is a very important relation in relativity indicating that a change in kinetic energy can be expressed in terms of mass as the variable. This equation is valid for change in energy in any form whatever.

Since the kinetic energy is zero when $v = 0$, then it is also zero when $m = m_0$, therefore integrating equation (25) we obtain

$$E_x = \int_0^{E_k} dE = c^2 \int_{m_0}^{m} dm = c^2\,(m - m_0) \qquad ...(26)$$

Thus the kinetic energy of a moving body equals c^2 times its gain in mass due to the motion. This relation suggests that we may think of the increase in energy as the actual cause of the increase in mass and implies that in addition to the familiar forms of energy, such as kinetic and potential etc., there is yet another kind 'mass energy'. According to this view equation (26) can be rewritten as

$$mc^2 = E_k + m_0 c^2. \quad \text{...(26a)}$$

The quantity m_0c^2 is called the *rest mass energy* of the particle. We can now define that total energy E of the particle as

$$E = mc^2 \quad \text{... (27)}$$

in which m is the relativistic mass of the particle. This equation leads to the idea of the *equivalence of mass and energy* and is certainly the most practical consequence of Einstein's theory of relativity.

Einstein's principle of relativistic mass energy equivalence may be stated as : "A mass m is equivalent to an amount of energy E and the equation relating these quantities is

$$E = mc^2,$$

where c is the speed of light".

This relation has received abundant experimental verification. It can be derived directly from the definition of the kinetic energy E_k of a moving body as the work done in bringing it from rest to its state of motion

i.e. $$E_x = \int_0^x F dx, \quad \text{...(28)}$$

where F is the component of the applied force in the direction of the displacement dx and x is the distance over which the force acts. Using the relativistic form of the second law of motion.

$$F = \frac{d(mv)}{dt}$$

Equation (28) becomes

$$E_x = \int_0^x \frac{d(mv)}{dt}\,dx$$

$$= \int_0^{mv} v\,d(mv)$$

$$= \int_0^v vd\left(\frac{m_6\, v}{\left(1-v^2/c^2\right)^{1/2}}\right) \qquad ...(29)$$

Since $m = \dfrac{m_9}{\left(1-v^2/c^2\right)^{1/2}}$ ...by Equation (22)

Integrating Eq. (29) by parts $\left(\int udw = uw - \int wdu\right)$

$$\therefore \quad E_k = \frac{m_0 V^2}{\sqrt{1-v^2/c^2}} - m_0 \int_0^v \frac{vdv}{\sqrt{1-v^2/c^2}}$$

$$= \frac{m_0 v^2}{\sqrt{1-v^2/c^2}} + m_0 \int_0^v \frac{-c^2\, vdv}{+c^2\sqrt{1-v^2/c^2}}$$

$$= \frac{m_0 v^2}{\sqrt{1-v^2/c^2}} + m_0\, c^2\left[\sqrt{1-v^2/c^2}\,\right]_0^v$$

$$= \frac{m_0 v^2}{\sqrt{1-v^2/c^2}} + m_0\, c^2\left[\sqrt{1-v^2/c^2} - 1\right]$$

$$= \frac{m_0 v^2}{\sqrt{1-v^2/c^2}} + m_0\, c^2\sqrt{1-v^2/c^2} - m_0 c^2$$

$$= \frac{m_0 v^2}{\sqrt{1-v^2/c^2}} + m_0\, c^2$$

$$= mc^2 - m_9 c^2$$

or $\quad mc^2 = E_k + m_0 c^2 \qquad ...(30)$

The quantity m_0c^2 is called the *rest-mass energy* of the particle, thus we can define the *total energy* of the particle as

$$E = mc^2 \qquad ...(31)$$

when m in the relativistic mass of the particle.

This equation leads us to the idea of *equivalence of mass and energy* and is known as Einstein's mass-energy equation.

This relation can also be expressed as :

$$E = \frac{m_0 v^2}{\sqrt{1 - v^2 / c^2}}. \qquad ...(32)$$

A very useful equation describing a relativistic particle is the one between the momentum and the total energy E giving us the invariant $E^2 - p^2c^2$. By equations (32) and (23)

$$E^2 - p^2c^2 = \frac{m_0^2 c^4}{1 - v^2 / c^2} - \frac{m_0^2 v^2 c^2}{1 - v^2 / c^2}$$

$$= m_0^2 c^4 \qquad \text{(constant)}$$

Expressing E in terms of momentum, we get

$$E = \left(m_0^2 c^4 + p^2 c^2\right)^{1/2}. \qquad ...(33)$$

This is the relativistic *energy momentum relation.*

Cosmic Ray Showers

These showers from outer space provide an other illustration of conversion of matter into energy. According to Millikan they are produced in some parts of the space where the temperature may be so low that the hydrogen nuclei may coalesce to form helium nucleus with liberation of energy corresponding to the *packing effect* in the form of radiation in accordance with Einstein's mass energy relationship, $E = mc^2$, where E is the energy liberated by a mass m and c is the velocity of light in vacuum. When four hydrogen nuclei unite to form helium nucleus, there is a loss of atomic weight equal to

$$1.0078 \times 4 - 4.002 = 0.0292$$

This loss of at. wt. measures liberation of energy

$$= 0.0292 \times 931 \times 10^6 \text{ eV}$$

$$[\because 1 \text{ at. wt.} = 931 \times 10^6 \text{ eV}]$$

$$= 27.1852 \times 10^6 \text{ eV}$$

The magnitude of the energy of cosmic rays gives likelihood to such a view.

Derivation of classical expression for E_k from relativistic expression. We can easily derive the classical expression and the kinetic energy from equation (26) which can also be written as

$$E_k = m_0c^2(K-1), \quad \text{...(34)}$$

since $m = km_0 \dfrac{m_0}{\left(1-\dfrac{v^2}{c^2}\right)^{\frac{1}{2}}}$ *cf.* equation (22)

This relativistic expression for the kinetic energy of a particle will reduce to the more familiar classical expression when $v<c$. This can be seen by expanding the first term K in the bracket of equation (34) by binomial theorem.

Now $K = \left(1-\dfrac{v^2}{c^2}\right)^{\frac{1}{2}} = 1 + \dfrac{1}{2}\dfrac{v^2}{c^2} + \ldots\ldots$

neglecting terms in v/c above the second power.

Equation (34) thus becomes

$$E_k = m_0c^2\left[1+\frac{1}{2}\frac{v^2}{c^2}+\ldots-1\right]$$

$$= \frac{1}{2} m_0v^2.$$

This is the classical expression for the kinetic energy of a particle.

Thus we see that the relativity theory does not "overthrow" the classical theory but rather extends and modifies it. We can say that, in general, classical mechanics constitutes approximate form of mechanical theory that is *valid for* any motion which is slow as compared with the speed of light. The need for the relativity theory in atomic and nuclear physics comes in part from the fact that fundamental particles such as electrons, protons and mesons etc. can travel with velocities approaching that of light with the result that kinematic and dynamical consequences of the Lorentz equations assume greater importance.

The four dimensional continuum. The special theory of relativity dealing with systems moving relatively to each other with uniform velocity and its application to four dimensional space is the intermediate stage in the evolution of the General Theory of Relativity which deals with accelerated systems and gravitational fields. Thus general theory makes a number of predictions namely— (1) motion of the perihelion of mercury, (2) deflection of a ray of light and (3) shift of spectral lines in an intense gravitational field and so on, which have since been verified substantially. It also shows that Newton's law of gravitation is an approximation of a more general law enunciated by Einstein, according to which gravitational attraction was shown to be a property of space. Being beyond the scope of this book, the general theory will not be dealt with here. However, the concept of four dimensional continuous may be elaborated a bit.

When the equation of a line in cartesian coordinates is transformed we arrive at the following results:

The distance s between two points with coordinates x_1, y_1, z_1, and x_2, y_2, z_2 in the frame F, say, is given by

$$s^2 = (x_2 - x_1)^2 + (y_2 - y_1)^2 + (z_2 - z_1)^2 \quad ...(35)$$

For the second frame of reference F' moving with respect

to the first with uniform velocity v, the application of the transformation already derived in equation (9), to the above equation yields

$$s^2 = K^2\{(x_2' - x_1') + v(t_2' - t_1')\}^2 + (y_2' - y_1')^2 + (z_2' - z_1')^2 \quad ...(36)$$

This clearly depends upon v and is certainly not similar in form to the original equation. The equation (35) therefore is not *invariant* as required by the first postulate and hence not applicable to the space time continuum. If, however, this equation is replaced by

$$s^2 = (x_2 - x_1)^2 + (y_2 - y_1)^2 + (z_2 - z_1)^2 - c^2(t_2 - t_1) \quad ...(37)$$

it will be found by substituting the transformation values of equation (9), that it *does remain invariant* (verify).

Therefore for a line element this invariant form of equation is

$$ds^2 = dx^2 + dy^2 + dz^2 - c^2dt^2 \quad ...(38)$$

If the velocity of light $c(=3 \times 10^8$ m/sec) is taken as unity for mathematical convenience, the invariant is slightly simplified to

$$ds^2 = dx^2 + dy^2 + dz^2 - dt^2 \quad ... (39)$$

Physically interpreted, it means that a *length is not described by three dimensions dx, dy, dz* but in addition, *the time dimension dt* is forced upon us.

The quantity *ds* is called *point-event* and does not correspond exactly to a distance in space but is an *interval in a four dimensional continuum in which space and time are closely linked together.*

It will be noticed from the first form of the invariant for a line element, equation (38), that when $ds = 0$, we have

$$c^2 = (dx^2 + dy^2 + dz^2)/dt^2 \quad ...(40)$$

i.e., something moving with the velocity of light. *This can only be a light photon* (or a ray of light), hence the equation $ds = 0$ is that for a ray of light.

In the four dimensional continuum the equivalent of a straight line is *a line making ds a minimum.* This is termed a *Geodesic.* The integral *ds, i.e.,* the expression $\int ds$, which denotes the *history* of a particle is called a *world line. The intersection of two world lines constitutes an ordinary 'event'.*

Some more examples of relativistic calculations :

Example. *Find the energy produced by the complete annihilation of one gram of matter using the mass energy equation* $E = mc^2$.

If m is expressed in grams, E in ergs, then c must be expressed in cm per sec. Taking $c = 3 \times 10^{10}$ cm s^{-1} a mass of 1 g is equivalent to energy given by

$$E = 1\,g \times (3 \times 10^{10}\text{cms}^{-1})^2$$

$$= 9 \times 10^{20}\text{ ergs.}$$

$$= \mathbf{9 \times 10^{18}\ joules} \quad [10^7\text{ ergs.}=1\text{ Joule}]$$

Similarly, a mass of 1 kilogram is equivalent to

$$E = 1\text{ kg} \times (3 \times 10^8\text{ms}^{-1})^2$$

$$= \mathbf{9 \times 10^{10}\ joules,}\text{ or } \mathbf{2.5 \times 10^{10}\ k\text{-}W\text{-}hr.}$$

Note. The basic unit of nuclear physics is the *atomic mass unit (amu)*, which is 13/12 of the rest mass of carbon 12.

One *amu* $=1.66 \times 10^{-27}$ kg and this is nearly the proton's mass. It is equivalent to

$$E = m_0c^2$$

$$= 1.66 \times 10^{-27}\text{ kg} \times (3 \times 10^8)^3\text{ m}^2\text{ s}^{-2}$$

$$= 1.49 \times 10^{-10}\text{ joule.}$$

In atomic and nuclear calculations, the erg or joule is an inconvenient unit of energy because the amounts of energy involved in single event usually are only small fractions of erg, say 10^{-4} to 10^{-12} erg. It is, therefore, the practice to express energies in units of electron-volts abbreviated as 'ev'. The electron volt is the energy acquired by any charged particle carrying a unit electronic charge, when it falls through a pot. diff. of 1 volt : it is equal to 1.6 × 10–12 erg or 1.6 × 10–12 joule. For conversion two other units are also used one equal to 10^3 ev called Kev and other equal to 10^6 ev called Mev.

1 *amu* may be expressed in ev and Mev.

$$1\ amu = \frac{\text{energy in jounles}}{\text{joules per ev}}$$

$$= 1.49 \times 10^{-10}\ \text{joule} \times \frac{1\ \text{ev}}{1.60 \times 10^{-19}\ \text{joule}}$$

$$= 9.31 \times 10^{8}\ \text{ev} = \mathbf{931\ Mev.}$$

Example. *A relativistic particle whose rest mass is m_0 is moving with a velocity of 0.9 c. Determine (a) its relativistic mass (b) its kinetic energy* (c) *why is it incorrect to use the expression $\frac{1}{2}mv^2$ for the kinetic energy of a relativistic particle ?*

The relativistic mass

$$m = \frac{m_0}{\left(1 - \frac{v^2}{c^2}\right)^{\frac{1}{2}}}$$

$$= \frac{m_0}{\left(1 - \frac{(0.9c)^2}{c^2}\right)^{\frac{1}{2}}} \quad ...[\because v = 0.9c]$$

$$= \frac{m_0}{(0.9c)^{\frac{1}{2}}} = \frac{m}{0.436} = \mathbf{2.3\ m_0}$$

(1) The kinetic energy

$$E_k = mc^2 - m_0c^2$$

$$= 2.3\ m_0c^2 - m_0c^2$$

$$= \mathbf{1.3\ m_0c^2.}$$

(*c*) Because we must take mass variation into account for a relativistic particle.

Example. *If an observer in motion moves away with a velocity c/2, how far will it be from the stationary observer in one second from the relativistic point of view ?*

According to classical mechanics in one second he will proceed a distance $c/2$. Relativistic equation 8(*a*) on the other hand gives

$$x_1' = K(x_1 - vt)$$

and $$x_2' = K(x_2 - vt)$$

$$\therefore \quad x_2' - x_1' = K(x_2 - x_1)$$

Let $X_2' - x_1'$ be the distance traversed in time t by the moving observer moving with a velocity $c/2$ as measured in its own frame, then,

$$x_2' - x_1' = ct/2$$

$\therefore$ distance as measured from the point of view of the stationary observer

$$x_2 - x_1 = \frac{x_2' - x_1'}{K} = \frac{ct/2}{\left(1 - \frac{v^2}{c^2}\right)^{-\frac{1}{2}}}$$

$$= \frac{ct}{2}\left(1 - \frac{v^2}{c^2}\right)^{\frac{1}{2}}$$

$$= \frac{ct}{2}\left(1 - \frac{v^2}{4c^2}\right)^{\frac{1}{2}} \qquad \ldots\left[\because v = \frac{c}{2}\right.$$

that is, in one sec he will be $\left(\frac{c}{2}\right)\left(1 - \frac{c^2}{4c^2}\right)^{\frac{1}{2}}$ or only 0.433 c away from the stationary observer's point of view.

Example. *Two particles in a given frame have velocities* u_{x1} *= —0.9 c and* u_{X2}*= +0.9 c. What is the velocity of one particle relative to the other ?*

Let us choose a reference frame moving with, say, the first particle. If this frame is called the *F* frame, then relative to this in the notation of equation 18 (*a*) the given frame has a velocity $v = -(-0.9c) = +0.9\ c$ and $u_{x_2} = w = +0.9e$.

Applying equation 18 (*a*) we obtain the velocity of the second particle relative to the first being equal to

$$u = \frac{v + w}{1 + \frac{vw}{c^2}} = \frac{+0.9c + 0.9c}{1 + (0.9c)0.9c) / c^2}$$

$$= \frac{+1.8c}{1 + 0.81} = \frac{1.80}{1.81}\ \text{c} = \mathbf{0.9994\ c}$$

According to classical mechanics the relative velocity

$$u = v + w = +0.9c + 0.9c = 1.8c.$$

This exceeds *c*.

The *relativistic* velocity, however, is always less than *c*.

Bibliography

Addicott, F.T. : *Optics,* University of California Press, London, 1982.

Atherton, J.C. : *Manipulation of Flowering,* Butterworths, London, 1987.

Ball. N.G., M.B. Wilkins : *The Physiology of Plant Growth and Development,* McGraw Hill book, London, 1969.

Bemier, G., J.M. Kinet : *Physical Optics,* C.R.C. Press, Boca Raton, 1981.

Benjamin, W.A. : California. Boyer, P.D. : *The Enzymes,* Academic Press, New York, 1970.

Bull, H.B. : *An Introduction to Physical Optics,* F.A. Davis Co. Philadelphia, Eisenburg, 1964.

Canny, M.J. : *Phloem Translocation,* Cambridge University Press, New York, 1973.

Cohen, P. : *Control of Enzyme Activity,* Chapman and Hall, London, 1976.

Crafts, A.S. and C.E. Crisp, : *Phloem Transport in Plants,* W.H. Freeman, San Francisco, 1973.

Douglas, J.S. : *Advanced Guide to Optics,* Pelham Books, London, 1976.

Epstein, E. : *Geometrical Optics,* John Wiley, New York, 1972.

Epstein, E. : *Mineral Nutrition of Plants : Principles and Perspectives,* John Wiley and Sons, New York, 1972.

Esau, K. : *Plant Anatomy,* John Wiley and Sons, New York, 1965.

Fershl, A. : *Enzyme Structure and Mechanism,* W.H. Freeman, San Francisco, 1977.

Forward, D.F. : *Energy and Carbon Metabolism,* Academic Press, New York. 1983.

Ftzler M.E. : *Plant Lectins : Molecular and Biological Aspects,* Academic Press, New York, 1985.

Gauch, H.G. : *Inorganic Plant Nutritionm,* Dowden, Hutchinson and Ross, Pennsylvania, 1972.

Govindjee, : *Bioenergetics of Photosynthesis,* Academic Press, New York, 1975.

Green, J.R. : *An Introduction to Vegetable Physiology,* London, 1911.

Halevy, A.H. : *Handbook of Optics,* C.R.C. Press, Boca Raton, 1985.

Loewy, A.G. and Siekevitz, P. : *Cell Structure and Function,* Rinehart and Winston, New York, 1969.

Mazliak, P. : *Light and Optics,* Chapmann and Hall, London, 1973.

McNeil, M. Darvill : *Structure and Function of the Primary Cell Walls of Plants,* McGraw, London, 1984.

Novikoff, A.B. : *The Endoplasmic Reticulum : a Cytochemist's View,* New York, 1976.

Ogawa, Y. and R.W. King : *Geometrical Optics,* Springer-Verlag, Berlin, 1990.

Opik, H. : *The Respiration of Higher Plants,* Edward Arnold Publishers, London, 1980.

Rabinowitch, E. and Govindjee, : *Photosynthesis.* John Wiley, New York, 1969.

Rains, D.W. J. Bortner and J.E. Varner : *Light,* Academic Press, New York, 1976.

Rubin, C. and Rosen, O. *Protein phosphorylation,* 1975.

Ruhland, W. : *Encyclopaedia of Optics,* Springer-Verlag, Berlin, 1961.

Salisbury, F.B. and C.W. Ross, : *Physical Optics,* Woodworth Publishing Co., Belmont, California, 1967.

Siegelman, H.W. and G. Hind : *Optics and Light,* Plenum, New York, 1978.

Srivastava, H.S. : *Elements of Physical Optics,* Rastogi Publications, Meerut, 1990.

Thimann, K.V. : *Optics,* CRC Press, Florida, 1980.

Tolbers P.C. : *Enzyme Kinetics,* Chapman and Hall, London, 1977.

Tolbert, N.E. : *Microbodies—Peroxisomes and Glyoxysomes,* Oxford University Press, London, 1971.

Vince, Pruce, : *Light and the Flowering Process,* Academic Press, London, 1984.

Walsh, C. : *Geometrical and Physical Optics,* Edward Arnold, London, 1979.

Weinheim, Osbome, D.J. : *Optics and Light,* Elsevier Biomedical Press, Holland, 1978.

Zeevart, J.A.D. : *Light and Optics,* McGraw Hill, London, 1976.

Zimmerman, M.H. and J.A. Milbum : *Encyclopaedia of Optics,* Springer Verlag, Berlin, 1975.